Human Genetics

Other books in the Biomedical Sciences Explained Series

0 7506 3256 9 Biochemistry *J.C Blackstock*
0 7506 2879 0 Biological Foundations *N. Lawes*
0 7506 3254 2 Biology of Disease *W. Gilmore*
0 7506 3111 2 Cellular Pathology *D.J. Cook*
0 7506 2878 2 Clinical Biochemistry *R. Luxton*
0 7506 2457 4 Haematology *C.J. Pallister*
0 7506 3413 8 Immunology *B.M. Hannigan*
0 7506 3253 4 Molecular Genetics *J. Hancock*
0 7506 3415 4 Transfusion Science *J. Overfield, M. Dawson and D. Hamer*

Acquisitions editor: Melanie Tait
Devlopment editor: Myriam Brearley
Production controller: Chris Jarvis
Desk editor: Jane Campbell
Cover designer: Helen Brockway

Human Genetics

Anne Gardner BSc DipRCPath
Principal Clinical Scientist (Molecular Genetics)

Rodney T. Howell BA FRCPath
Consultant Clinical Scientist (Cytogenetics)

Teresa Davies BSc PhD FRCPath CertMHS
Consultant Clinical Scientist (Cytogenetics)

Regional Cytogenetics Centre, Southmead Hospital, Bristol, UK

Series Editor:
C.J. Pallister PhD MSc FIBMS CBiol MIBiol CHSM
Principal Lecturer in Haematology, Department of Biological and Biomedical Sciences,
University of the West of England, Bristol, UK

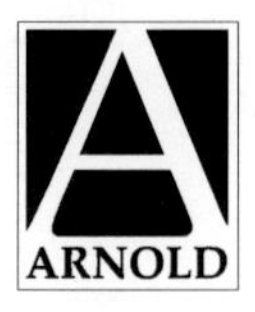

A member of the Hodder Headline Group
LONDON
Co-published in the United States of America by
Oxford University Press Inc., New York

First published in Great Britain in 2000 by
Arnold, a member of the Hodder Headline Group,
338 Euston Road, London NW1 3BH

http://www.arnoldpublishers.com

Co-published in the USA by
Oxford University Press Inc.,
198 Madison Avenue, New York, NY 10016
Oxford is a registered trademark of Oxford University Press

British Library Cataloguing in Publication Data
A catalogue record for this book is available from the British Library

Library of Congress Cataloging-in-Publication Data
A catalog record for this book is available from the Library of Congress

ISBN 0 340 76374 4

1 2 3 4 5 6 7 8 9 10

Typeset by David Gregson Associates, Beccles, Suffolk
Printed and bound in Great Britain by The Bath Press, Avon

What do you think about this book? Or any other Arnold title?
Please send your comments to feedback.arnold@hodder.co.uk

Contents

Preface

Most people nowadays have heard of genes, chromosomes and DNA, either from television documentaries or science fiction films. While they may not automatically associate these terms with the concept of 'genetics', there is now a public perception of how modern discoveries in this field may impact on our lives, in a way unimaginable just a few years ago.

Genetics began to be recognized as a distinct scientific discipline around the beginning of the twentieth century, following the rediscovery of Mendel's earlier publications concerning the mechanisms of inheritance in peas, and the growing acceptance of the concepts contained in Darwin's 'The Origin of Species'.

It is hard to believe that as late as 1956, we did not even know the correct number of chromosomes found in the nuclei of human cells; it was thought that humans had 48 chromosomes until Tjio and Levan determined the correct number to be 46.

Only three years before had the structure of DNA been determined, yet today scientists routinely sequence large lengths of human DNA, and in the process appear to discover new genes (and therefore the cause of more human genetic diseases) on practically a daily basis!

Why then did there appear to be such a long delay and apparent lack of interest in human genetics in the first half of the twentieth century? Why do so many theories and discoveries seem to be based on bacteria, the fruit fly *Drosophila*, or mice?

The answers lie in the fact that different strategies are needed to study human disease. Bacteria are easily grown in the laboratory on agar plates under controlled conditions, and because they can divide and grow so rapidly, any change in their DNA (a mutation) can be rapidly detected. Similarly fruit flies or even mice have a fast reproductive cycle so that the scientist can choose particular matings to suit his or her experiment. In addition, because these animals have been studied for many years, much is already known about their DNA, genes and chromosomes (which are often fewer in number and more simply organized than ours).

Humans have a long reproductive cycle of nine months; it is therefore very difficult to assemble pedigrees (family trees) of many generations. If an interesting genetic disorder arose in a particular family, it could only be observed: there is no equivalent of the controlled matings used in laboratory animals.

Gradually, however, such observations began to reveal various ways in which human genetics could be studied. When mutations in critical regions of DNA occur they affect the normal functioning of genes, which in turn may fail to produce an important gene product such as an enzyme or a hormone. This may occasionally result in a recognizable disorder, i.e. a genetic disease which may appear throughout a family pedigree over several generations.

This leads to the concept of the genetic syndrome, in which a particular chromosomal abnormality or DNA mutation always results in the affected individual having a more or less consistent pattern of features including defects in organs, recognizable facial features and typical behavioural characteristics.

This book is intended to take the reader through the basics of human genetics by studying how our DNA, genes and chromosomes work, and what happens when they go wrong. The chapters are logically organized into larger sections, starting with 'Mechanisms of Disease', followed by 'Diagnosis of Disease' and finally 'Prevention of Disease'.

Thousands of genetic disorders have now been characterized; some of these are described in other books in the *Biomedical Sciences Explained* series. For example, inborn errors of metabolism are biochemical disorders, while haemoglobinopathies are diseases affecting the blood. To this end only those genetic diseases which are now modern paradigms are discussed in depth in this book, such as those relating to genomic imprinting, triplet repeat disorders and tumour suppressor genes. The Appendix provides a quick reference for conditions described in less detail in the text.

Although the book is aimed at level 1 and 2 undergraduates – medical students, other trainee professionals and anyone with an interest in genetics may find it helpful as an introduction to this exciting and rapidly developing field. We all have a right to understand the genetic forces that affect our families and descendants, and to know how far we have come in the battle to predict and prevent genetic disease.

A. Gardner, R.T. Howell and T. Davies

Series preface

The many disciplines that constitute the field of Biomedical Sciences have long provided excitement and challenge both for practitioners and for those who lead their education. This has never been truer than now as we face the challenges of a new millennium. The exponential growth in biomedical enquiry and knowledge seen in recent years has been mirrored in the education and training of biomedical scientists. The burgeoning of modular BSc (Hons) Biomedical Sciences degrees and the adoption of graduate-only entry by the Institute of Biomedical Sciences and the Council for Professions Supplementary to Medicine have been important drivers of change.

The broad range of subject matter encompassed by the Biomedical Sciences has led to the design of modular BSc (Hons) Biomedical Sciences degrees that facilitate wider undergraduate choice and permit some degree of specialization. There is a much greater emphasis on self-directed learning and understanding of learning outcomes than hitherto.

Against this background, the large, expensive standard texts designed for single subject specialization over the duration of the degree and beyond, are much less useful for the modern student of biomedical sciences. Instead, there is a clear need for a series of short, affordable, introductory texts, which assume little prior knowledge and which are written in an accessible style. The *Biomedical Sciences Explained* series is specifically designed to meet this need.

Each book in the series is designed to meet the needs of a level 1 or 2 student and will have the following distinctive features:

- written by experienced academics in the biomedical sciences in a student-friendly and accessible style, with the trend towards student-centred and life-long learning firmly in mind;
- each chapter opens with a set of defined learning objectives and closes with self-assessment questions which check that the learning objectives have been met;
- aids to understanding such as potted histories of important scientists, descriptions of seminal experiments and background information appear as sideboxes;
- extensively illustrated with line diagrams, charts and tables wherever appropriate;
- use of unnecessary jargon is avoided. New terms are explained, either in the text or as sideboxes;
- written in an explanatory rather than a didactic style, emphasizing conceptual understanding rather than rote learning.

I sincerely hope that you find these books as helpful in your studies as they have been designed to be. Good luck and have fun!

C.J. Pallister

Acknowledgements

Slides of the children depicted in the plates were provided by Dr Peter Lunt with kind permission from their parents. Thanks are due to Dr Martin Davies for Figures 1.1–1.4 and Figures 9.1, 9.2, 9.3 and 9.5, to Ann Oliver for the ultrasound scan and to Mary Gable for her perspective and critical appraisal at the draft stage.

Part One:

Mechanisms of Disease

Chapter 1

Structure and organization of DNA

Learning objectives

After studying this chapter you should confidently be able to:

Outline the levels of secondary structure of DNA leading to the chromosome.

Describe the structure of the DNA molecule.

Outline the principles of DNA replication.

List the phases of the cell cycle.

Describe the stages of mitosis and meiosis and explain the resultant numbers of chromosomes.

Define exons and introns with respect to a typical gene.

Give examples of repetitive DNA.

Identify the band position on a chromosome using the ISCN.

Our bodies consist of around nine billion cells. Most of our cells contain a nucleus with its nuclear membrane, surrounded by cytoplasm. Within the cytoplasm are found various structures known as **cell organelles.**

The nucleus contains the hereditary information stored in the form of **deoxyribonucleic acid** (**DNA**). This is known as **nuclear DNA.**

Although the structure and functions of the cytoplasmic organelles are too many to list individually, the mitochondria (singular mitochondrion) are of interest as they have their own DNA, known as **mitochondrial DNA.**

Eukaryotes, which include such diverse organisms as yeasts and humans, are characterized by DNA contained within a nuclear membrane, and the presence of cell organelles.

Prokaryotes have no nuclear membrane, they have no introns in their genes, and their gene controlling systems are organized to respond to rapid changes in the environment. DNA and RNA are found together. Eukaryotes have more complex control systems. Eukaryotic RNA is transported from the nucleus to the cytoplasm.

A nucleosome comprises eight protein molecules called **histones**. These are basic proteins comprising four pairs called H2A, H2B, H3 and H4 respectively. In its partially folded state, DNA is bound to a fifth histone, H1, thought to participate in the control of gene expression by tighter binding of the DNA to prevent transcription.

Prokaryotes, for example bacteria, have no clear division between nucleus and cytoplasm and may only have one simple 'chromosome' called a nucleoid.

When a cell is not dividing, it is in a state known as **interphase.** The DNA is in a very decondensed state. From the point at which a cell begins to divide however, the DNA begins to condense into the highly coiled shapes known as **chromosomes.**

The nuclei of body (**somatic**) cells have 46 chromosomes comprising 22 pairs of autosomes and one pair of sex chromosomes, whereas in eggs or sperm (the **germ** cells) there are 23 chromosomes. When fertilization occurs the correct number of chromosomes is restored.

If a chromosome is gradually unravelled, it can be seen that the DNA is supercoiled into a solenoid-like form, which in turn is composed of thinner fibres. As these are unwound it can be seen that the DNA is wrapped around protein structures called **nucleosomes** (Figure 1.1).

If the histone proteins of the nucleosome are removed, the basic form of DNA known as the **double helix** is revealed, comprising two separate strands wound around each other such that, when viewed from above, they normally appear to spiral in a clockwise or right-handed direction (Figure 1.2).

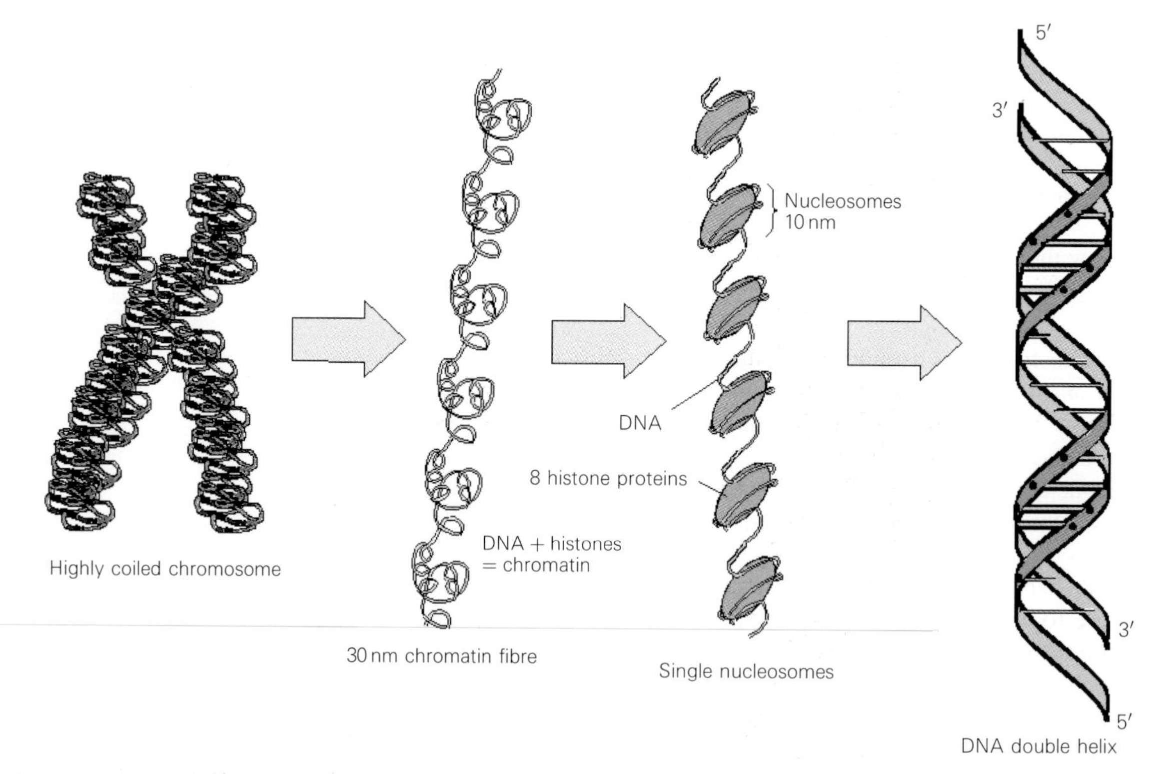

Figure 1.1 *Uncoiling a chromosome to reveal the DNA double helix*

The main spiral of the double helix is known as the **sugar–phosphate backbone**. The phosphate is attached to the outer part of a five-sided sugar 'ring' – the deoxyribose molecule. Attached to the inner part of the sugar ring is one of four different nitrogenous bases. The bases are arranged towards the centre of the two single helixes, and lie almost flat, in a stacked formation (Figure 1.3).

Two **bases** (each comprising two organic rings), adenine (**A**) and guanine (**G**), are called **purines**. The bases cytosine (**C**) and thymine (**T**) (each comprising one organic ring) are called **pyrimidines** (Figure 1.4).

A human diploid cell nucleus (i.e. 46 chromosomes) contains about two metres of DNA helix.

Sometimes a cytosine may be modified by the addition of a methyl (CH_3) group attached to the fifth atom of the carbon ring. The cytosine is then said to be methylated.

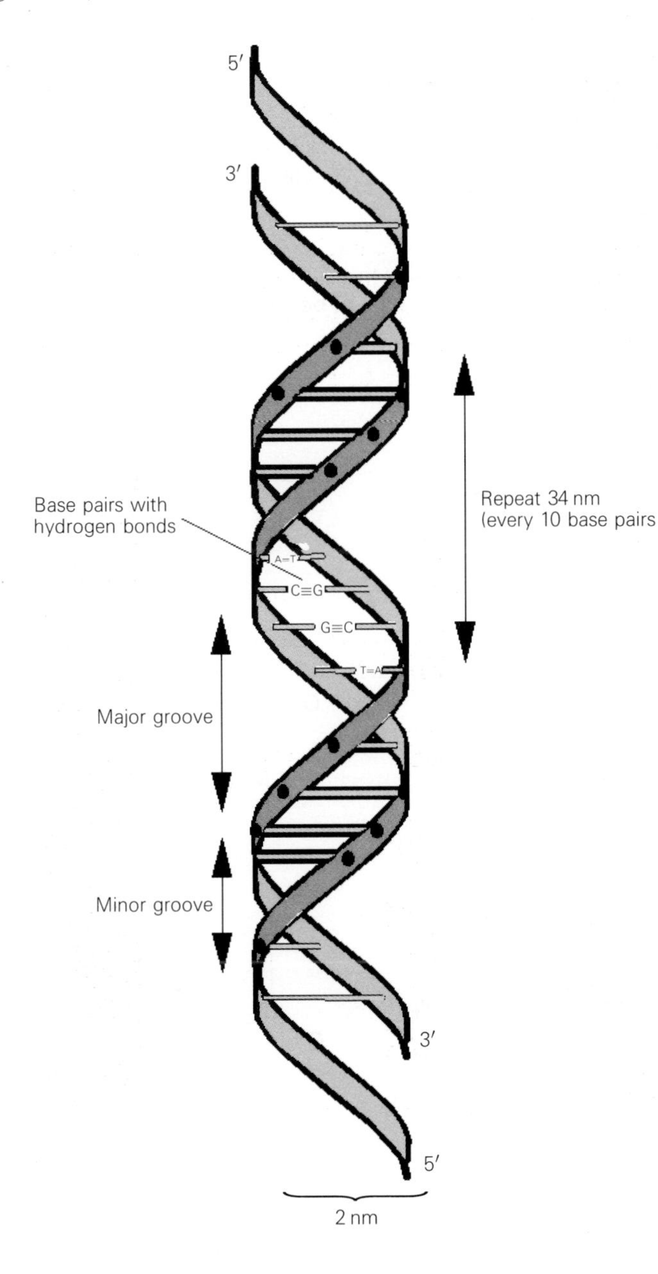

Figure 1.2 *The DNA double helix*

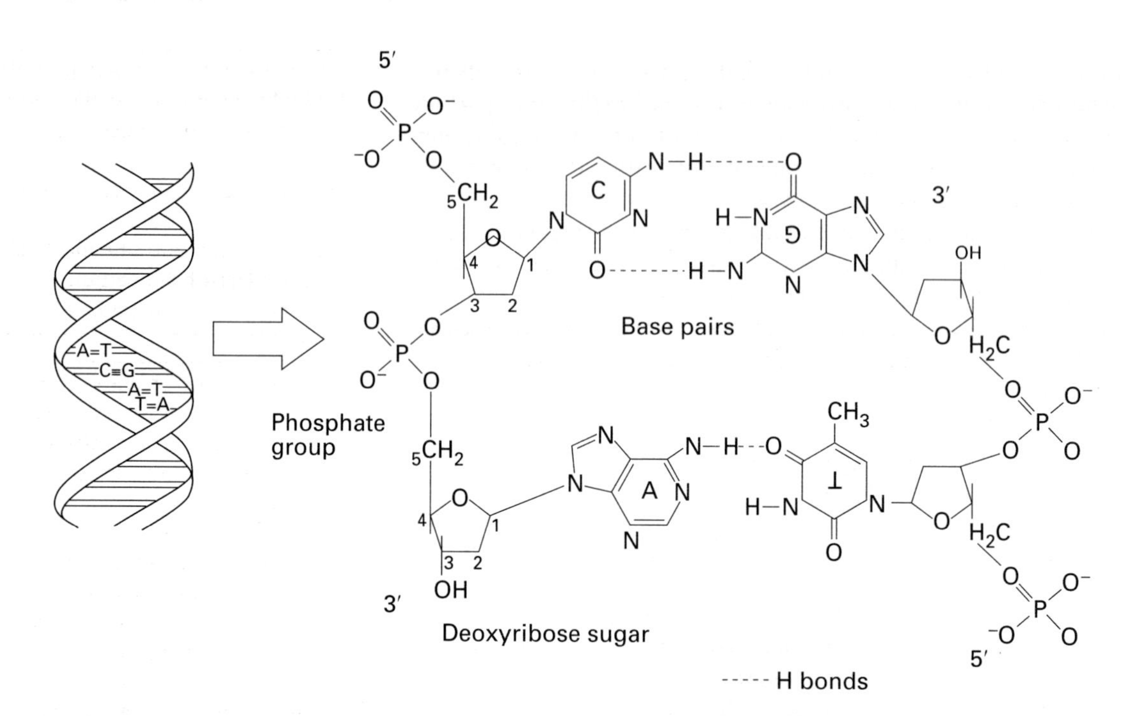

Figure 1.3 *Detailed structure of the DNA helix*

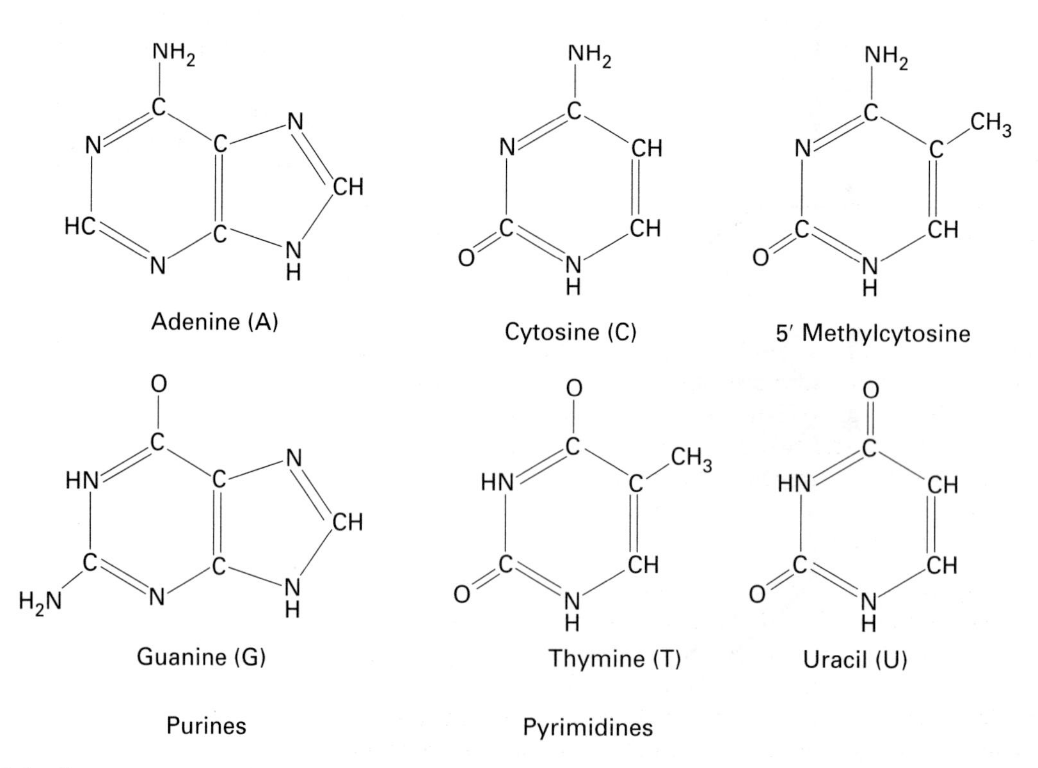

Figure 1.4 *Bases*

Organic rings may contain carbon, nitrogen and/or oxygen atoms. The carbons are consecutively numbered in a clockwise direction around the ring of the pentose sugar. It is then possible to describe how various molecules may be attached to each other at a particular carbon position on a ring.

In Figure 1.3 the phosphate group can be seen attached to the 5′ (five **prime**) carbon of the deoxyribose, whereas the base is attached to the 1′ carbon.

One chain of the double helix has a 5′ phosphate at one end and a 3′ OH group at the other end. The other chain lies **antiparallel** to the first, as it has the 3′ OH group at the top, next to the 5′ of the first chain. This helix will end in a 5′ phosphate. This polarity is important in DNA replication and in the action of certain modifying enzymes used in DNA techniques.

The double helix is held together with hydrogen bonds formed between pairs of bases. A always pairs with T, using two hydrogen bonds, whereas C pairs with G more strongly, as there are three hydrogen bonds. This is known as **complementarity**.

If two DNA strands are separated (denatured), as can be achieved by heat or an alkaline pH, although the hydrogen bonds are broken the strands may reassociate again if the temperature is lowered or the pH is returned to neutral. This phenomenon is sometimes referred to as **reannealing** or, if it is part of a molecular biology technique, **hybridization**.

Because each strand of a DNA helix is **complementary** to the other, any faulty or missing bases may be replaced, as the opposing base (or sequence of bases) will provide the guidelines for repair.

The sugar ring of messenger RNA (mRNA) is a ribose molecule with a hydroxyl (OH) group attached at the 2′ carbon, instead of a hydrogen as in the deoxyribose of DNA.

mRNA has the base **uracil (U)**, whereas DNA has the base thymine (T).

The sugar together with the base is known as a nucleoside; the sugar, base and triphosphate group is a deoxynucleoside triphosphate (**dNTP**). An example is deoxycytidine triphosphate, written dCTP.

A **nucleotide** is the basic subunit of DNA, comprising the sugar, base and interconnecting phosphate group. DNA is therefore a polymer of nucleotide units.

The pairing of an A with a T, or a C with a G results in what is known as a **base pair**, written as 'bp', and:

1000 bp = 1 kilobase (kb)
1000 kb = 1 Megabase (Mb)

DNA replication

DNA replication is possible because the sequence of bases on one strand of the double helix is complementary to that on the other.

Principle of replication

The double helix has to unwind for replication to take place. Each pre-existing (**conserved**) strand serves as a pattern or **template** for the new complementary strand. Once copying has taken place there will then be four strands: two old strands and two new strands.

DNA replication is therefore said to be **semi-conservative.**

Eukaryotic replication

In the eukaryotic cell all the DNA is copied during the **S (synthesis)** phase of a continuing **cell cycle** (discussed on the following pages).

Eukaryotic DNA synthesis has more than one start point for replication. The replicating units are called **replicons**.

At the localized site of a replicon, a helicase enzyme unwinds the DNA helix. Replication occurs in both directions at a **replication fork** (Figure 1.5).

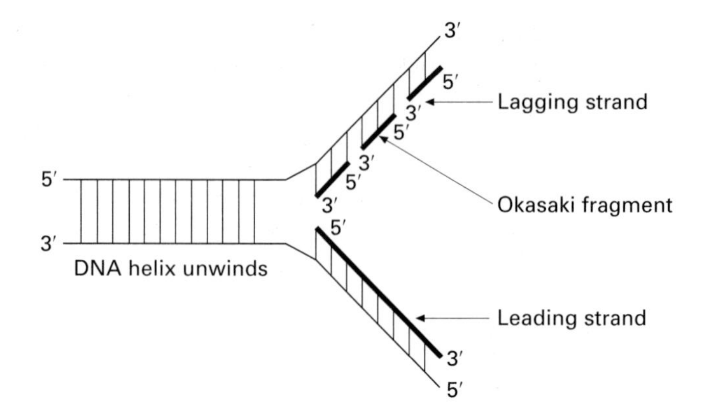

Figure 1.5 *DNA replication*

A prokaryote such as a bacterium has one circular chromosome with a single point of origin for the start of replication, which proceeds in two directions until there are two rings.

Replication proceeds continuously in the 5′ to 3′ direction, with nucleotides attached to the 3′ end. This **continuously formed DNA strand** is called the **leading strand.**

In order that the DNA can be copied from the 3′ to 5′ direction, **short pieces** of around 1000 bases called **Okasaki fragments** are synthesized in the 5′ to 3′ direction; these are then joined together by a ligase.

This DNA sequence is called the **lagging strand.**

DNA synthesis is finally completed when all the individual replicons meet.

There are many enzymes involved in the manipulation of DNA.

- A **polymerase** copies DNA or RNA by incorporation of the appropriate complementary base (actually a deoxynucleoside triphosphate – a dNTP). Some DNA polymerases can also proof-read by spotting mistakes and removing the wrongly synthesized bases.
- An **exonuclease** removes dNTPs.
- An **endonuclease** nicks a strand of DNA.
- A **ligase** joins together (ligates) two pieces of double-stranded DNA at their sugar–phosphate (phosphodiester) bonds.

The cell cycle

A somatic cell exists in either a resting stage, a growing stage or a state of cell division, when it divides into two daughter cells. The cell cycle then begins again.

The process of cell division is called **mitosis**; at all other stages the cell is in **interphase.**

Interphase is divided into several other phases (Figure 1.6):

- G_0 is a resting phase from which cells move into G_1.
- G_1 ('Gap 1') is a growing phase which, in actively growing cells, lasts on average around 16–24 hours, during which the cell makes various molecules necessary for the following stages.
- **S phase** is when DNA synthesis occurs. This lasts on average 8–10 hours.
- G_2 ('Gap 2') lasts for about four hours, during which there is protein synthesis and membrane assembly. At the end of this stage the cell is committed to mitosis, and the DNA starts to condense into chromosomes.
- **Mitosis** is the stage at which cell division occurs. This lasts around one hour and is itself split into several phases. This process is discussed in more detail later in this chapter.

G_1 has an extremely variable timespan, depending on the growth and division requirements of the particular cell type. Rapidly growing cells such as embryonic cells may omit G_1 altogether.

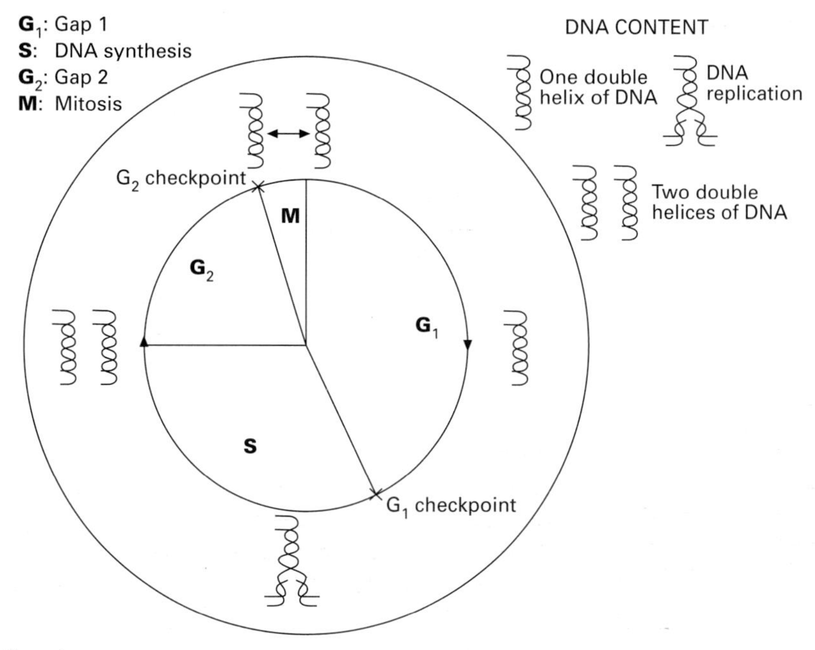

Figure 1.6 *The cell cycle*

Mitosis and meiosis

Mitosis

Mitosis is the stage at which somatic cell division occurs to produce two identical daughter cells. As a result of DNA replication during S phase of the cell cycle, each chromosome in a cell entering mitosis will consist of two identical strands known as **chromatids.**

Mitosis occupies only about one hour of the whole cell cycle. It is a continuous process but is usually described as divided into four stages: prophase, metaphase, anaphase and telophase.

- **Prophase** begins when the chromosomes become visible following condensation. Each chromatid consists of a pair of long thin parallel strands or sister chromatids which are held together at the **centromere.**
- The nuclear membrane disappears and the **nucleolus** (the site of ribosomal RNA synthesis) becomes undetectable as its component parts disappear. The **centriole** divides and its two products migrate towards opposite poles of the cell.
- **Metaphase** begins when the chromosomes have reached their maximal contraction. They move to the equatorial plate of the cell and the spindle forms. The spindle consists of microtubules, formed by the centrioles and consisting of the protein **tubulin.** They connect the centrioles to the **kinetochore** of the centromere for each chromosome. The two chromatids of each chromosome begin to separate until they are only connected at the centromeric region.

Progression around the cell cycle is controlled by molecules called **cyclins** which direct other associated molecules called **cyclin dependent kinases (CDKs)**. These enzymes add phosphate groups on to certain gene products to regulate their activity. This is known as **phosphorylation**.

There are two important points in the cell cycle that commit the cell to go on to the next stage.

The first is the **G_1 to S checkpoint**, when the cell checks that there is no DNA damage before replication, and that all the correct genes for replication are turned on.

The second is the **G_2 to S checkpoint**, when the cell checks that all the correct apparatus for chromosome segregation has been synthesized and is present.

Mitosis occurs in all embryonic tissues and continues at a lower rate in most adult tissues other than in end cells (e.g. neurones) and is vital for both tissue formation and maintenance.

- **Anaphase** begins when the centromeres divide and the paired chromatids separate, each to become a daughter chromosome. They move, centromere-first, to the poles of the cell.
- **Telophase** starts when the daughter chromosomes reach each pole of the cell. The cytoplasm divides, the cell plate forms and the chromosomes start to unwind. The nuclear membrane reforms at this stage. The two cells can now enter the cell cycle again at G_1 or enter the resting phase G_0.

It is during the later stages of prophase and during metaphase that the chromosomes can be most readily examined for abnormalities by cytogeneticists.

Meiosis

Gametogenesis, the production of gametes, occurs in the gonads. The somatic diploid chromosomal complement ($2n$) is halved to the haploid number (n) of a mature gamete. This is done in such a way as to ensure that each gamete contains one member of each pair of chromosomes. This reduction is achieved by meiotic cell division. Fusion of the egg and sperm restores the diploid number.

Meiosis consists of one DNA replication but two successive divisions: the first and second meiotic divisions.

First meiotic division (MI)

This consists of the stages prophase, metaphase and anaphase. Chromosomes at this point consist of two chromatids. Homologous chromosomes pair, side by side, with the exception of the X and Y chromosomes in male meiosis in which only the ends pair. Exchange of homologous segments occurs between chromatids from each of the pair of homologous chromosomes – this is known as **crossing over** or **recombination.** During prophase I in the male the X and Y chromosomes pair in the pseudoautosomal region or PAR (see Chapter 2) and there is a single obligatory crossover.

Prophase of the first meiotic division is complex and five stages can be recognized:

- leptotene
- zygotene
- pachytene
- diplotene
- diakenesis.

Since the chromosomes assort independently during meiosis, this results in 2^{23} or 8 388 608 different possible combinations of chromosomes in the gamete from each parent. Hence there are 2^{46} possible combinations in the zygote. There is still further scope for **variation** provided by crossing over during meiosis. This can result in combinations of genes on a chromosome different from those on the chromosomes in the parent. If there is, on average, only one crossover per chromosome (there are likely to be more than this) and a 10% paternal/ maternal allele difference, then the number of possible zygotes exceeds 6×10^{43}. This number is greater than the number of human beings who have so far existed!

Leptotene starts with the first appearance of the chromosomes. At this stage each chromosome consists of a pair of thread-like sister chromatids.

During **zygotene** (synapsis) homologous chromosomes pair up, starting at the telomeres and proceeding towards the centromeres, to form **bivalents.** These are closely bound together by the **synaptonemal complex.** In the male the paired X and Y chromosomes form a sex bivalent. This is condensed early in pachytene as the **sex vesicle.**

Pachytene is the main stage of chromosomal thickening or condensation. Each chromosome is now seen to consist of two chromatids, and hence each bivalent is a **tetrad** of four strands. Crossing over occurs, during which homologous regions of DNA are exchanged between chromatids.

During **diplotene** the bivalents start to separate. Although the two chromosomes of each bivalent separate, the centromere of each remains intact. The two chromatids of each chromosome remain together. During the longitudinal separation the two members of each bivalent can be seen to be in contact at several places, called **chiasmata** (singular chiasma; see Chapter 4). These mark the location of crossovers, where the chromatids of homologous chromosomes have exchanged material in late pachytene. On average, there are about 52 chiasmata per cell in the human male and more in the female. At diplotene in males the sex bivalent opens out and the X and Y chromosomes can be seen attached to one another.

Diakenesis is the final stage of prophase, during which the chromosomes coil more tightly.

Metaphase begins when the nuclear membrane disappears and the chromosomes move to the equatorial plane. At anaphase the two members of each bivalent disjoin, one going to each pole. These bivalents are assorted independently to each pole. The cytoplasm divides and each cell now has 23 chromosomes, each of which is a pair of chromatids, differing from one another only as a result of crossing over.

Spermatogenesis occurs in the seminiferous tubules of the male from the time of sexual maturity onwards. At the periphery of the tubule are **spermatogonia**, some of which are self-renewing stem cells, with others already committed to sperm formation. The **primary spermatocyte** is derived from a committed spermatogonium. The primary spermatocyte undergoes the first meiotic division to produce two **secondary spermatocytes** each with 23 chromosomes. These cells undergo the second meiotic division, each forming two **spermatids**. The spermatids mature without further division into **sperm**.

The production of mature sperm from a committed spermatogonium takes about 60–65 days. Normal semen contains 50–100 million sperm per millilitre. Sperm production continues (albeit at a reduced rate) into old age. The numerous replications that occur during sperm production are thought to increase the chance of mutation (see Chapter 3). It has been shown that the risk for several single gene mutations is increased in the offspring of older men.

Second meiotic division (MII)

The second meiotic division follows the first without an interphase. It resembles mitosis and has the stages of prophase, metaphase, anaphase and telophase; the centromeres now divide and sister chromatids pass to opposite poles.

Meiosis has three important consequences:

- Gametes contain only one representative of each homologous pair of chromosomes.
- There is random assortment of paternal and maternal homologues.
- Crossing over ensures uniqueness by further increasing genetic variation.

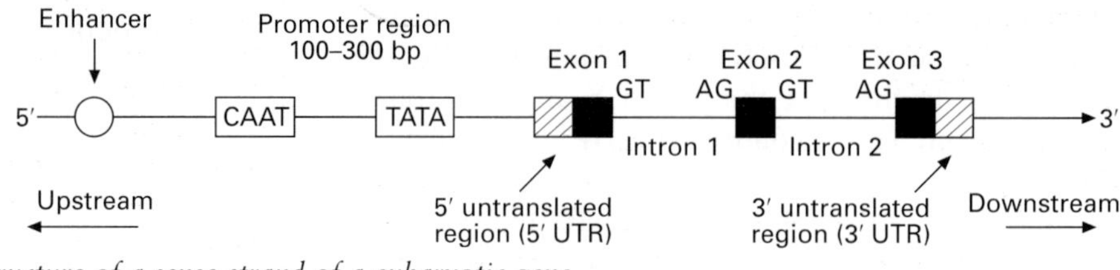

Figure 1.7 *Structure of a sense strand of a eukaryotic gene*

In contrast to spermatogenesis, the process of **oogenesis** in the female is largely complete at birth. **Oogonia** are derived from the primordial germ cells. By about the third month of foetal life the oogonia have become **primary oocytes** and some of these have already entered the prophase of first meiosis. The primary oocyte remains in a phase of maturation arrest in prophase I (unlike spermatocytes) until sexual maturity is reached. This stage is known as **dictyotene**. As each individual follicle matures and releases its oocyte into the Fallopian tube the first meiotic division is completed. Hence completion of the first meiotic division in the female may take over 40 years.

The first meiotic division results in an unequal division of the cytoplasm, with the **secondary oocyte** receiving the great majority of the cytoplasm and the daughter cell known as the **first polar body** consisting largely of only a nucleus.

Meiosis II then commences, during which fertilization can occur. This second meiotic division results in the formation of a **second polar body**. It is

Genes

The nuclear genome

The nuclear genome consists of the DNA arranged in genes contained within chromosomes in the cell nucleus. There are thought to be 85–100 000 genes in the human genome.

Eukaryotic genes comprise **exons** and **introns** (Figure 1.7). The exons are the important coding sequences; any change in this code (**a mutation**) may lead to an abnormal protein being produced.

Exons are consecutively numbered from the 5′ end of the gene as exon 1, exon 2 and so on.

In between genes, and indeed in between exons of genes, are the intervening sequences (IVS) or introns. These stretches of DNA appear to contain no useful coded information. Although this DNA may originally have coded for genes, these no longer function, and so a mutation in an intron would be unlikely to cause a genetic disorder.

A particular fragment of DNA is not necessarily identical in different people. Benign differences are known as **polymorphisms.** As we will see in Chapter 7 on Molecular genetics, these polymorphisms can be exploited by molecular biologists to study genetic diseases in families because they are inherited in a Mendelian fashion.

It has been suggested that 80–90% of the human genome comprises non-coding stretches of DNA, and that 40–50% of this is repetitive in nature. There are varying degrees of repetitiveness, as shown in Table 1.1. Repetitive human DNA is also called satellite DNA.

Gene size in humans may range from around 100 base pairs for tyrosine transfer RNA to the moderately large gene for cystic fibrosis (250 kb). One of the largest human genes known is that coding for the protein dystrophin which is dysfunctional in Duchenne muscular dystrophy. This gene is 2.4 Mb long, yet is still too small to be seen at the level of resolution of a light microscope.

The direction to the left of the 5′ end of a gene is said to be **upstream,** whereas the direction to the right of the 3′ end of a gene is **downstream** (Figure 1.7).

Table 1.1 *Repetitiveness at the DNA level*

Name of DNA sequences	*Comments*
Unique	Single copy DNA. There is about 10% of coding DNA in the whole human genome.
Microsatellite	Also known as short tandem repeats (STRs), 1–4 bp e.g. CA, TA, tetranucleotides. Found on all chromosomes.
Minisatellite	Variable number of tandem repeats (VNTRs), 9–64 bp. These were the original 'fingerprinting' probes. Found on all chromosomes especially near telomeres. Telomeres, 6 bp, TTAGGG, up to several kb. Found at the ends of all chromosomes.
Interspersed	SINES (short interspersed nuclear elements). An example is the human Alu repeat of about 280 bp, up to one million copies mostly in the euchromatic (G light) areas of chromosomes.
Interspersed	LINES (long interspersed nuclear elements). The L-1 or Kpn repeat is 1.4–6.1 kb, up to 100 000 copies, mostly in the heterochromatic (G dark) areas of chromosomes.
Satellite	Alphoid (α) Repeat, 171 bp: up to several Mb in length. Found in centromeric heterochromatin of all chromosomes.
Satellite	Beta (β) Repeat, 68 bp: 100 kb to several Mb in length. Found in centromeric heterochromatin of chromosomes 1, 9, 13, 14, 15, 21, 22 and Y.
Satellite 1	Repeat 25–48 bp. A–T rich. Found in the centromeric heterochromatin of most chromosomes, and other heterochromatic regions.
Satellite 2	Repeat 5 bp. Probably found on all chromosomes.
Satellite 3	Repeat 5 bp. Probably found on all chromosomes.
Triplet repeats	Repeats of 3 bp, usually CGG, CAG, CCG or CTG. Tens of copies in the normal person can expand to thousands in an affected individual. Unique in that these dynamic mutations may expand from genera tion to generation and actually cause certain genetic syndromes. See also Chapters 3 and 7.

thought that the long interval between the onset of meiosis and its completion in the female contributes to the increase in risk of failure of homologous chromosomes to separate during meiosis (non-disjunction) in the older mother. This is associated with an increased risk of trisomy for chromosome 21 (Down syndrome) and for other chromosomes (see Chapter 5).

Mutations resulting in the production of abnormal proteins form the fundamental basis of many genetic diseases.

The sex determining region gene (*SRY*) and the human mitochondrial genome do not have any introns.

The mitochondrial genome

Mitochondria have their own circular genome comprising mitochondrial DNA (mtDNA). The genome is around 16.5 kb in size.

Two circular strands known as the H and L strands code for 13 proteins involved in energy production and 22 **transfer RNA (tRNA)** molecules, together with two ribosomal RNAs (rRNA).

Some proteins are produced by a combination of nuclear genes and mitochondrial genes. The nuclear products are transported through the cytoplasm to the mitochondria where they associate with mitochondrial gene products to form a functional protein. An example is the cytochrome c oxidase complex, which comprises three subunits coded by mitochondrial genes and 10 subunits coded by nuclear genes.

Chromosomes and chromatin

The structural organization of DNA does not stop at the DNA level. Within the chromosome are found specialized categories of chromatin.

Heterochromatin and euchromatin

Heterochromatin comprises highly condensed chromatin fibres. There are two types of heterochromatin: constitutive heterochromatin and facultative heterochromatin.

Constitutive heterochromatin is composed of repetitive DNA sequences containing **no transcriptionally active genes**. It replicates late in the cell cycle at the end of the S phase. Its function is unknown, but it may have a role in conserving and protecting vital genes from recombination. In human chromosomes, most of this heterochromatin is located close to the centromeres.

Facultative heterochromatin can be **transcriptionally active or inactive**, and is also late replicating. It is found in the selectively inactivated X chromosome in females.

Euchromatin contains chromosome fibres which are less densely packed than heterochromatin. Euchromatin is generally **transcriptionally active**, but may contain some regions of transcriptionally inactive DNA.

Nucleolar organizers or **NORs**, the genes controlling ribosomal RNA synthesis, occur in repeated blocks on the short arms of chromosomes 13, 14, 15, 21 and 22. These genes have to be active until mitosis and therefore remain uncoiled while the remainder of the mitotic chromosome is condensed. The NORs are visible as secondary constrictions.

Chromosomes and nomenclature

Until the early 1970s, chromosomes were only identifiable by shape and size, as they were just stained and observed by light microscopy as solid blocks of colour. From 1971, the technique of G-banding ensured that each chromosome could be differentiated from others by virtue of the individual band patterns.

For the purposes of analysis, the 22 autosomes and the pair of sex chromosomes are arranged in a **karyotype** according to their size and group (Figure 1.8).

By convention, autosomes are numbered in descending order of size, from 1 to 22. The short arm is designated 'p' and the long arm 'q'. Chromosomes with the centromere close to the mid-point are called **metacentric**, those with a noticeably shorter short arm are **submetacentric**, and those with a very short short arm are **acrocentric**. Some other species, mice for example, have **telocentric** chromosomes where the centromere is located right at the end – a shape not seen in normal human chromosomes.

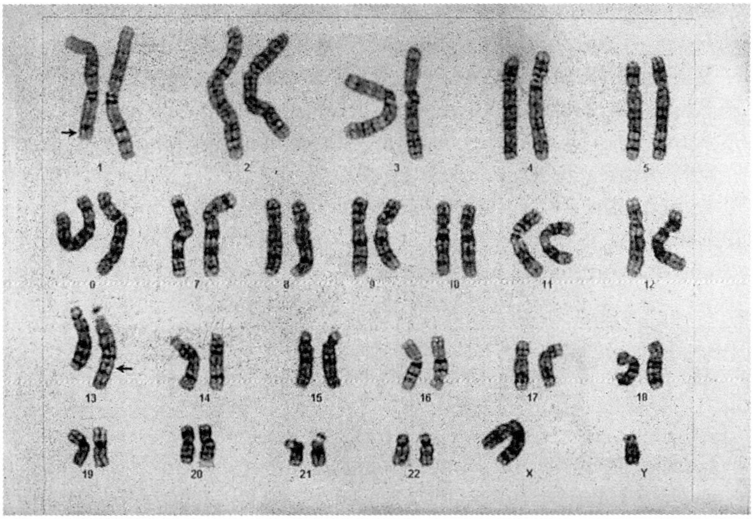

Figure 1.8 *The karyotype of a man, referred for chromosome analysis because of his wife's recurrent miscarriages, showing a balanced reciprocal translocation between the long arm of chromosome 1 and the long arm of chromosome 13. The ISCN nomenclature is 46,XY,t(1;13)(q32;q33)*

As well as numbering each individual chromosome, they are sometimes classified as groups A to G based on their general size and centromere position, so that A comprises 1–3, large and metacentric, B 4–5, large submetacentric, C 6–12, medium-sized submetacentric, D 13–15, medium-sized acrocentric, E 16–18, smaller submetacentric, F 19–20 small metacentric and G 21–22, small acrocentric.

The International System for Human Cytogenetic Nomenclature, or ISCN, divides the banding pattern of each chromosome arm into regions and bands so that every band can be uniquely identified by a simple shorthand method. For example, 1p35 refers to chromosome 1, short arm, region 3, band 5. The lowest region and band numbers are closest (or **proximal**) to the centromere, and the highest numbers are at the tips, or **distal**.

The karyotype and chromosome analysis are discussed further in Chapter 5.

Some abbreviations used when writing karyotypes in ISCN:

del	deletion
der	derivative
dic	dicentric
dup	duplication
fra	fragile site
h	heterochromatin
i	isochromosome
ins	insertion
inv	inversion
mat	maternal origin
p	short arm
q	long arm
t	translocation

A semicolon (;) separates altered chromosomes and breakpoints in structural rearrangements involving more than one chromosome while a comma (,) separates chromosome numbers, sex chromosomes, and chromosome abnormalities.

Suggested further reading

Gardner, E.J. and Snustad, D.P. (1984). *Principles of Genetics*, 7th Edn. Wiley.

Strachan, T. and Read, A.P. (1999). *Human Molecular Genetics 2*, Ch. 1. Bios Scientific Publishers Ltd.

Self-assessment questions

1. Where is DNA located in the cell?
2. Outline the structure of DNA.

3. Define complementarity and write down the complementary bases to the following sequence: AGGTTCGGAT.
4. Why are exons of genes more important than introns?
5. Name three classes of repetitive DNA and give examples.
6. Name the four stages of the cell cycle. How long does a cell spend in each stage?
7. Explain the key differences between meiosis and mitosis.
8. Using the ISCN, write down the shorthand for the long arm of chromosome 2, band 3, region 5.

Key Concepts and Facts

Location and Arrangement of Hereditary Information

- DNA is the hereditary information stored in the cell nucleus. In association with histone proteins, it is condensed into 46 chromosomes.

Internal Structure of DNA

- DNA is an antiparallel double helix, comprising an outer sugar–phosphate backbone and four internal bases held together by hydrogen bonding in A:T and G:C base pairs. This ensures that one DNA strand is complementary to the other, and can therefore be replicated accurately.

The Cell Cycle

- At interphase the cell cycle is divided into four phases referred to as G_0, G_1, G_2 and S. Mitosis (M) is the stage at which cell division occurs.

Cell Division

- There are two kinds of cell division: mitosis and meiosis. Mitosis is somatic cell division and results in 46 chromosomes per cell, while meiosis occurs in germ cells and results in 23 chromosomes per cell.

Organization of Genomic and Repetitive DNA

- Genes comprise conserved coding exons and polymorphic non-coding introns. Repetitive DNA is also present in large quantities.

The Normal Karyotype

- The 46 human chromosomes are arranged pictorially in pairs in descending size order, with the sex chromosomes at the end. Banding enables each pair to be differentiated from any other pair.

Chapter 2

How normal genes work

Learning objectives

After studying this chapter you should confidently be able to:

Describe briefly the mode of action of structural and controlling genes.

Outline the processes of transcription and translation.

Explain the basic concept of imprinting and uniparental disomy.

Give examples of different levels of gene control.

Describe the three basic methods of DNA repair.

Outline the importance of somatic recombination and allelic exclusion in immunogenetics.

Name and describe the function of three major classes of developmental genes.

Explain the importance of the *SRY* gene with respect to sexual development.

When Crick and Watson elucidated the structure of DNA in 1953, they realized that complementary base pairing could provide a model for both DNA replication and the inheritance of genetic material.

During the following years, the steps by which the DNA sequences of genes were 'read' and turned into a string of amino acids to make up a polypeptide (a subunit of protein) were also determined.

DNA has to be **transcribed** into an intermediate molecule, RNA, which is then **translated** into amino acids, the correct sequences of which give rise to an active protein. In eukaryotes, DNA is found in the cell nucleus. However, RNA, amino acids and hence proteins are generally found in the cytoplasm.

It was also quickly established that particular genes coded for certain proteins which made up physical parts of the human body such as keratin found in hair and nails, or the coloured pigment found in eyes or skin. These are **structural genes**.

Gene control may occur directly or indirectly via a **signal transduction pathway**. Signals from outside the cell are directed into the cytoplasm and also into the nucleus.

James Watson had the following central dogma pinned up on his wall at Cambridge: **DNA to RNA to protein**.

By convention, diagrams of transcription always display the 5′ end of RNA on the left, which implies that it was copied from the antisense DNA strand.

The splicing out of introns is so precise that a nuclear intron will always start with GT and end with AG.

There are other transcriptional differences between prokaryotes and eukaryotes in that eukaryotic mRNA undergoes two further steps in transcription:

- A chemical called 7-methyl-guanosine is added to the 5′ end of the mRNA – this is known as **capping**.
- A long series of adenine bases (As) is added to the 3′ end – a **poly(A) tail**.

The production of proteins from a gene is called **gene expression.**

Some genes may not encode for an obvious end product – their proteins or RNA control other genes by switching them on and off. These are **controlling genes**. This chapter will look at some mechanisms by which controlling genes influence gene expression.

Transcription

The double-stranded DNA must unwind from its helix to be copied by an enzyme called RNA polymerase II. The antisense DNA strand (the one that runs 3′–5′) is copied such that the resulting RNA reads 5′–3′. This will then have exactly the same sequence as the DNA 5′–3′ sense strand, except that the thymines have been replaced by uracils (Figure 2.1).

In eukaryotes the newly transcribed RNA will include sequences for introns as well as exons (sometimes called **primary transcript RNA**). As this primary RNA passes from the nucleus into the cytoplasm, the introns loop out, and are carefully excised. The exons are then **spliced** together (Figure 2.2). Prokaryotic RNA does not need splicing as there are no introns. Human RNA without introns is called **messenger RNA** or **mRNA.**

Translation

This is the process whereby the mRNA is **translated** into protein. The cytoplasm contains structures called ribosomes, which themselves are made up of protein subunits derived from **ribosomal RNA (rRNA)**. rRNA is mainly produced in the **nucleolar organizer regions (NORs)** of the nucleolus (within the nucleus). NORs are found at the satellites of acrocentric chromosomes.

The mRNA passes through the cytoplasm and proceeds to the ribosomes (Figure 2.3) where the sequence is read in groups of three bases (**triplets**) called **codons**. Apart from three 'stop codons' (see below), each triplet corresponds to a particular amino acid. As

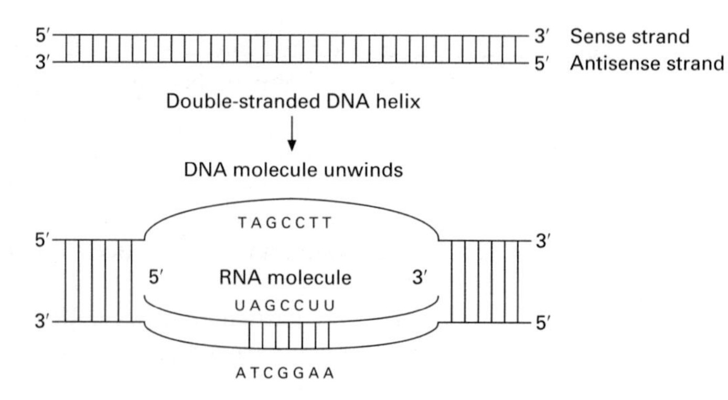

Figure 2.1 *Principles of transcription from DNA to RNA*

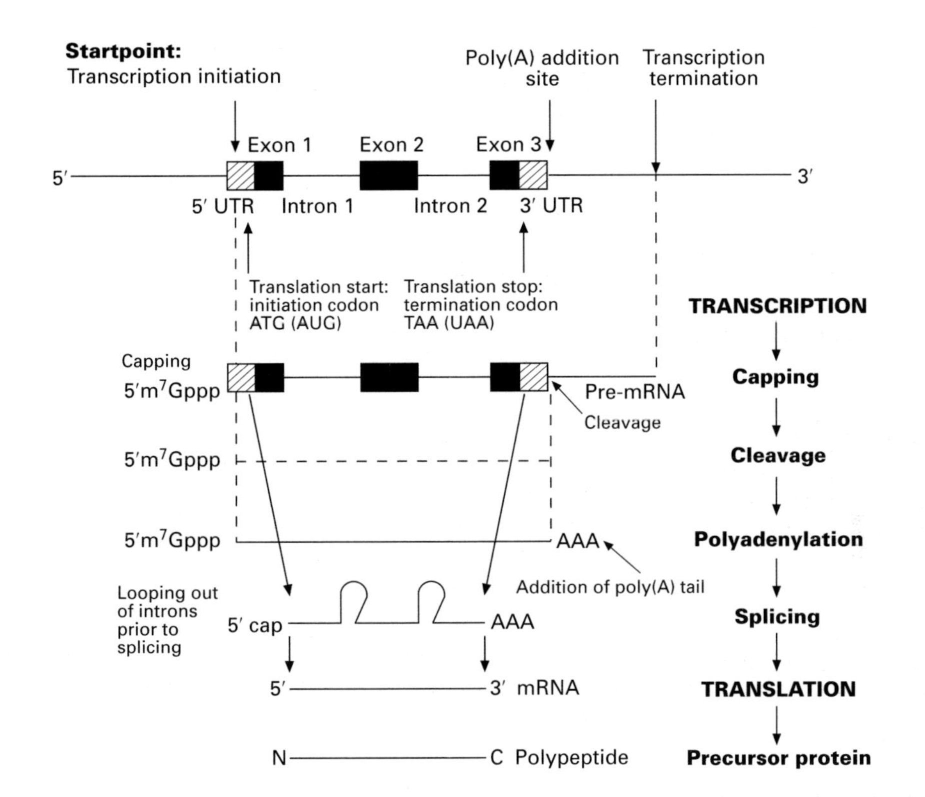

Figure 2.2 *Transcription and translation*

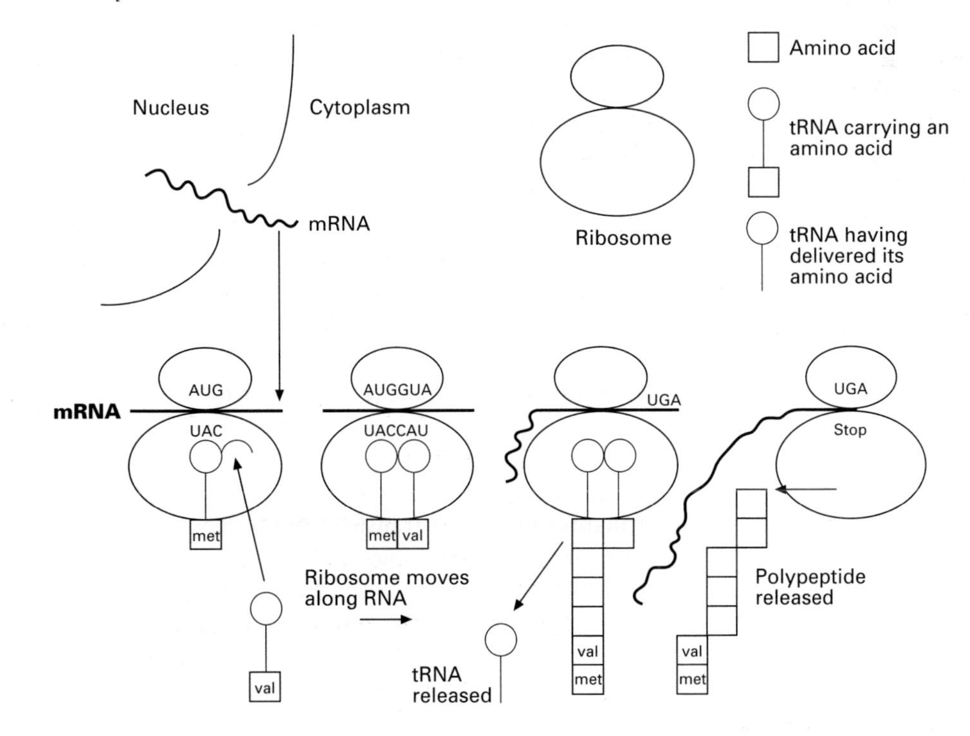

Figure 2.3 *Translation of mRNA by ribosomes and tRNA molecules*

The mitochondria do not use all the same triplet codons as nuclear DNA, as shown by the examples below.

Codon:
UGA
AUA
AGA

Nuclear genome:
STOP
Ile
Arg

Mitochondrial genome:
Trp
Met
STOP

Table 2.1 *The degenerate genetic code*

	First base				
Second base	*U*	*C*	*A*	*G*	*Third base*
U	UUU F UUC F UUA L UUG L	CUU L CUC L CUA L CUG L	AUU I AUC I AUA I AUG M	GUU V GUC V GUA V GUG V	U
C	UCU S UCC S UCA S UCG S	CCU P CCC P CCA P CCG P	ACU T ACC T ACA T ACG T	GCU A GCC A GCA A GCG A	C
A	UAU Y UAC Y UAA X UAG X	CAU H CAC H CAA Q CAG Q	AAU N AAC N AAA K AAG K	GAU D GAC D GAA E GAG E	A
G	UGU C UGC C UGA X UGG W	CGU R CGC R CGA R CGG R	AGU S AGC S AGA R AGG R	GGU G GGC G GGA G GGG G	G

Key to amino acids:

A	alanine	I	isoleucine	R	arginine
C	cysteine	K	lysine	S	serine
D	aspartic acid	L	leucine	T	threonine
E	glutamic acid	M	methionine	V	valine
F	phenylalanine	N	asparagine	W	tryptophan
G	glycine	P	proline	Y	tyrosine
H	histidine	Q	glutamine	X	stop codon

there are four bases (**A C G and T**), there are 64 (4 × 4 × 4) possible **codes** – but not all are used, as there are only around 20 amino acids (see Table 2.1). As several triplets can code for the same amino acid, this is a **degenerate** or redundant code. However, the **initiation codon** is almost always **AUG** (which uniquely codes for **methionine**) and is recognized as such by the ribosome.

As the triplet codes are read, the correct amino acid has to be placed in sequence, almost like beads on a string. This is done by structures known as **transfer RNAs** (tRNAs), whose appearance in some organisms resembles a cloverleaf.

At the 'stem' end is attached a particular amino acid – for example valine. On the opposing 'leaf' end is the **anticodon** for that amino acid.

The mRNA codon (GUA) will be complementary to the tRNA anticodon sequence (CAU) and thus the correct amino acid (V) will be delivered to the growing amino acid chain (Figure 2.3).

Many polypeptide chains can be synthesized simultaneously in the cytoplasm by the attachment of several ribosomes to the same RNA molecule.

Table 2.2 *Levels of genetic control*

Stage	*Level*	*Examples*
Transcriptional	Molecular: general	Promoters, transcription factors
	Molecular: specific	Enhancers, silencers
	Large scale: general	Methylation
		Chromatin structure
Post-transcriptional		Alternative splicing
Translational	Molecular	Translational response elements
Other		Dosage
		X-inactivation
		Genomic imprinting

Three stop codons, UAA, UAG or UGA, signal chain termination. The polypeptide chain is then released from the ribosome.

Most genes have to be periodically switched on or off, and are regulated to produce their proteins appropriately. An example is the production of insulin in the pancreatic cells as a response to eating sugar, i.e. expression in the correct place at the correct time.

Control of gene expression

Most genes are not **constitutively expressed**, i.e. they are not turned on all the time and in all cells. Those that do are obviously very important and are constantly required for the correct day to day functioning of the organism's cells – indeed, they are actually called **housekeeping genes.**

Eukaryotic gene expression is usually regulated during transcription or translation, either at the molecular level or on a larger scale (see Table 2.2 for a summary).

In **bacteria**, one gene can control several structural genes, which may be expressed sequentially in response to (for example) a nutrient appearing in the environment, when appropriate digestive enzymes may be produced. Jacob and Monod called this arrangement an **operon**.

Levels of eukaryotic gene control

Transcriptional control

The basis of molecular control

A gene can be defined as **one gene codes for one polypeptide** (or protein); in eukaryotes such a gene sequence is also called a **cistron.**

The untranslated region at the beginning of a gene is called the 5′UTR (**un**t**r**anslated **r**egion); at the end of the gene is the 3′UTR.

Promoters and transcription factors

As genes consist of both exons and introns, then the logical start position for transcription should be at the beginning (5′) of exon 1, terminating at the 3′ end of the last exon. In fact these areas are flanked by pieces of untranslated DNA; the actual initiation and termination areas are shown in Figure 2.2.

Just before the initiation site (at about 25 bases 5′) is often found a consistent run of bases, TATAA, known as a **TATA box.** Some genes may have another box known as the **CAAT box** with the

When **DNA sequences** are **conserved** because they have an important function, they are called **consensus sequences**. An example would be the TATA box. Sometimes amino acids are conserved in the same way, as their three-dimensional structure may be essential to the function of the protein. All these groups are known as **motifs**.

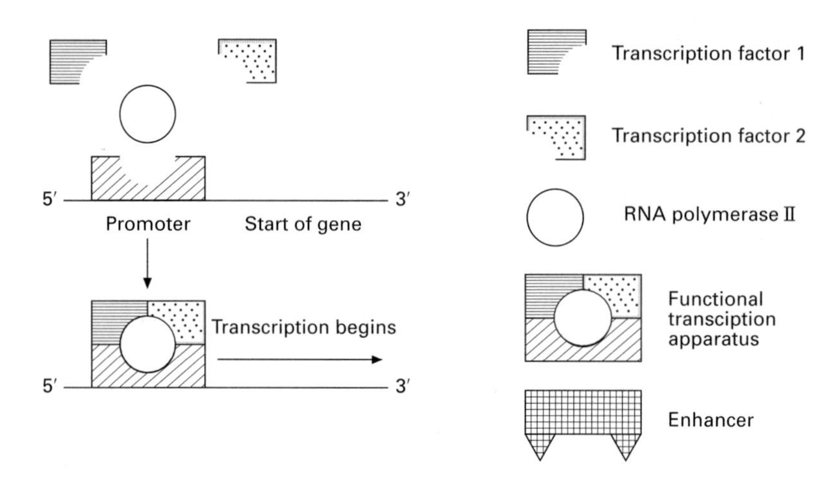

Despite being several kb away, an enhancer can switch on a gene by completing the transcription apparatus and activating the promoter

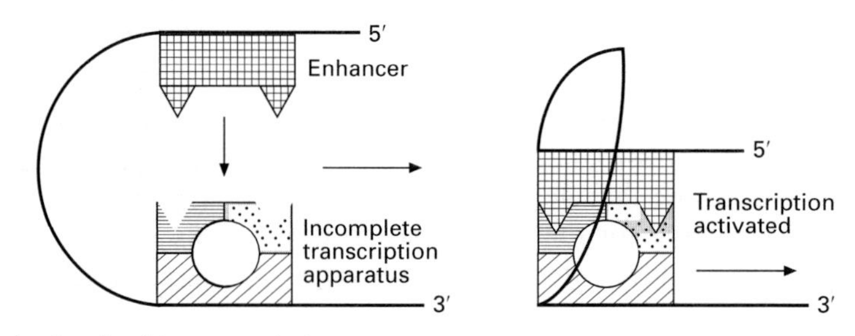

Figure 2.4 *Molecules involved in transcription*

Transcription factors have different regions which take on specific conformations to fit their target DNA and the RNA polymerase **II** like a jigsaw piece. They have odd sounding names like the zinc finger motif, the leucine zipper and the helix-loop-helix.

Enhancers are interesting because they may lie several kilobases away from their target gene and its promoter. A mutation in an enhancer may cause a clinical disorder due to lack of a particular protein, yet the actual gene seems undamaged.

sequence CCAAT. This is equivalent to an '**on**' switch, and is known as the **promoter** region.

For transcription to begin, RNA polymerase II (itself a small protein) has to 'log on' to the beginning of the DNA molecule in order to transcribe the sequence into RNA. For this purpose small proteins known as **transcription factors** are employed to assist the RNA polymerase II.

A specific transcription factor (TFIID) binds to the promoter region first, then other transcription factors guide the RNA polymerase to its correct position, which enables transcription to be turned on at a basic level. This is the transcription unit (Figure 2.4).

Specific transcriptional control

Sometimes transcription unit formation does not reach maximum efficiency; this is achieved by certain sequences called **enhancers** (Figure 2.4) which optimize transcription of a gene.

The opposite of enhancers are **silencers** (or repressors) which switch genes **off**.

Methylation

The DNA base **cytosine** can be **methylated** by the addition of a methyl group (CH_3) (see Figure 1.4). Methylation therefore tends to be found in **CG**-rich sequences.

Methylation is often seen in vertebrate DNA and generally follows the rule that **active** genes (i.e. those that are transcribing) have **less** or **no methylation. Inactive** (or repressed) genes show **more methylation** with **no gene expression**, as transcription factors may not recognize the DNA sequences.

However it does not follow that all gene expression is regulated by methylation, as the non-vertebrate fruit fly *Drosophila* does not have methylated DNA at all.

At the 5′ ends of actively transcribing genes are found short fragments of DNA called **CpG islands**. Although not every gene has these CpG islands, they are thought to represent the **beginning of housekeeping** (or other widely expressed) **genes**.

Chromatin structure

Chromatin structure consists of DNA and its associated histones together with other basic and acidic proteins. The primary coils of DNA are organized into higher levels of very condensed structures (see Chapter 1).

The nature of the chromatin structure itself is therefore important in gene expression. An open chromatin structure free of methylation will generally indicate that those genes can become transcriptionally active. There will be room for the correct transcription factors and appropriate enzymes to bind properly.

Areas of condensed chromatin (heterochromatin) signal suppression or inactivation of genes. There will be certain areas of constitutive heterochromatin (e.g. the centromeres) or telomeric chromatin which are inactive a great deal of the time, as these domains have an important structural function in terms of holding the chromosome together properly.

Any active genes which are inadvertently moved near to these inactive areas run the risk of being inactivated themselves. We will look at abnormal **position effects** in the next chapter.

It becomes apparent that genes retain specific functions partly because of the three-dimensional chromatin position in which they are naturally found, yet other factors such as methylation also play an important role. It is possible, however, for whole chromosomes (or areas of chromosomes) to become inactivated, resulting in essentially haploid areas of the human genome. This is not abnormal; functional haploidy probably arose during evolutionary

Sometimes gene expression may be a response to an external factor, such as the presence of a hormone inducing transcription. Genes that respond in this way contain **response elements**.

It is also possible for a gene to have more than one promoter which may produce different versions of the same protein in different body tissues (for example, muscle dystrophin may be a different length to brain dystrophin).

In a CpG doublet the 5′ **c**ytosine is connected by a **p**hosphodiester bond to a 3′ **g**uanine. This way of writing bases therefore represents the DNA sequence read 'horizontally' and not 'vertically' as for a C:G base pair.

CpG islands can be detected using the methylation sensitive restriction enzyme Hpa II (which cannot cut methylated DNA). The resulting small pieces of cleaved DNA must therefore be unmethylated and are sometimes called **HTF** islands (**H**pa **T**iny **F**ragments).

Telomeres cap the ends of eukaryotic chromosomes and maintain chromosomal stability. The human telomere consists of a stretch of tandemly repeated DNA sequences comprising six nucleotides (TTAGGG), together with a telomere-binding protein which protects the terminal DNA. If telomeres are lost, the end of the DNA helix becomes unstable and may fuse with the ends of other broken chromosomes or be subject to degradation.

Unmodified mRNA is unstable. As it passes into the cytoplasm it is quickly degraded by RNAses. One way translational control is expressed is by protection of mRNA, in order that efficiency of translation of particular mRNAs is maximized.

Protection may be achieved using long 5′ and 3′ untranslated regions, by the formation of secondary structures such as stems and loops, or by generation of hundreds of copies of important mRNAs which are protected from degradation by the ribosomes themselves.

In a similar manner to transcriptional response elements, the presence of certain substances may elicit a translational response. For example, when a particular molecule required for the production of haemoglobin (haem co-factor) is present, the rate of translation of globin RNA in the reticulocytes increases.

Autosomal genes are present in pairs – one on each homologous chromosome. If the level of protein required for the cell to function normally requires that **both genes should be active**, they will exhibit **bi-allelic expression**.

In some circumstances (due, for example, to the deletion of one of a pair of alleles) only 50% of the protein will be produced from the remaining intact gene. A haploid level of protein may not be enough for the cell to behave appropriately, and it therefore displays **haploinsufficiency** (see Chapter 3).

development to balance and protect the human genome (see also later sections on dosage and genomic imprinting).

Post-transcriptional regulation

After transcription of the primary RNA, the point at which the introns are spliced out gives another opportunity for differential gene expression. It is possible to splice out certain exons as well as introns; this produces different mRNAs and is known as **alternative splicing.**

Translational control

The operon arrangement of the bacterial genome enables a rapid response to its environment. Translational control in humans fulfils a similar function, as examples usually arise when a rapid response is required.

Because different cell types have different mRNAs in their cytoplasm, translational control must be rare. If it were not rare, we would see all possible mRNAs in every cell type.

When an egg is fertilized by a sperm, there is a large increase in the rate of protein synthesis, yet this does *not* require the production of new mRNAs after fertilization. This means that there must be pre-existing stable mRNAs in the mother's unfertilized egg which are only translated after fertilization.

Other methods of gene control

Dosage

At first glance, we would look at a human nucleus with 46 chromosomes and say that 23 had to be inherited from each parent and must therefore have equal importance in that cell.

This cannot be true, however, because females have two X chromosomes, whilst males only inherit one X from their mother (together with a Y from their father). Without **dosage compensation** females would have twice the amount of gene product as males with only one X. One of the X chromosomes is therefore **inactivated.**

X chromosome inactivation

One of the two Xs in female cells is **randomly inactivated** at the early blastula stage in embryonic development. Females are therefore a mosaic (mixture) of cells containing maternally **or** paternally inactive X chromosomes in a ratio of approximately 1:1. The parental origin of inactivation is now preserved as a particular somatic cell divides.

Control of X-inactivation

X-inactivation is controlled by a gene called ***XIST*** (pronounced 'exist') found in the **X-inactivation centre (XIC)** on the long arm of the X chromosome at Xq13. Although most of the genes on the inactivated X will not be expressed, ***XIST* is expressed in females**, as it has to be **active** to initiate the inactivation of its own X chromosome.

Although *XIST* is responsible for the initiation and spreading of X-inactivation, the XIC is not needed for the maintenance of X chromosome inactivation, as the XIC can be lost but the genes still remain silent due to methylation.

What is certain, however, is that the **single X in the male has to remain active**, and to do this it is necessary that *XIST* is switched **off**. Another way of expressing this is to say that the *XIST* gene on the male X is **imprinted.**

There are certain small areas which are left active on the inactivated X. One of these is the **pseudoautosomal region** (PAR) on the tip of the short arm at Xp22.3, where there are important **sequences shared with the Y chromosome** at Yp11.3. This is the area where an **obligatory crossover occurs at meiosis.**

It is thought that around 20 genes remain active on the inactive X; as they are not all in the PAR (some are on the long arm) this suggests that the inactivation proceeds individually in small increments (perhaps gene by gene).

Genomic imprinting

It is logical that dosage compensation exists on the X chromosome, in order that both males and females only express one X allele. It might be supposed that autosomes have no requirement for a dosage mechanism, as both males and females receive 22 autosomes from each parent.

Genomic imprinting results in one of two otherwise identical alleles being marked (imprinted) during gametogenesis such that the **homologous alleles are expressed differently depending on the parent of origin.**

In a female nucleus the inactive X can be seen as a distinctive densely staining structure called the Barr body. Barr described its composition as 'sex chromatin' – it is highly condensed heterochromatin.

Resetting of the imprint in germ cells

Imprinting is found in somatic cells. When an individual reproduces, however, any imprint received from their parents has to be **reset**; it is wiped out, so that in the chromosomes passed on to their own children, the new imprint is from the correct parental sex (Figure 2.5).

Not all Xs are inactivated in the same way as in humans. In marsupials such as kangaroos the paternal X is always inactivated, whilst in some male scale insects the whole paternal chromosome set is inactivated!

This is not comparable to dosage in humans: the paternal X or paternal genome in these animals is imprinted (see imprinting) and therefore not functionally equivalent to the female X or the female genome.

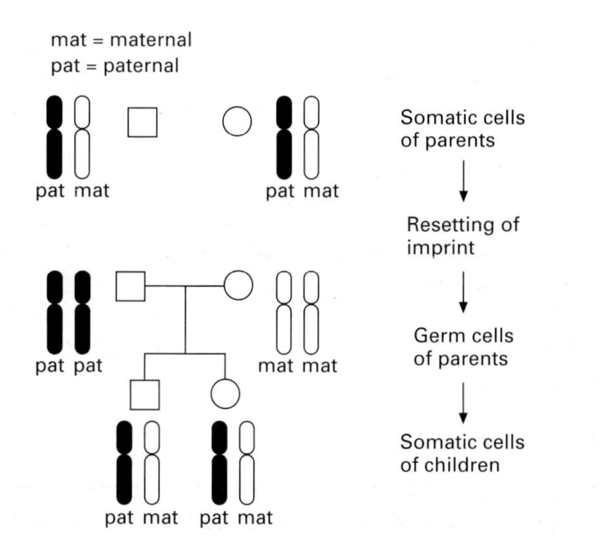

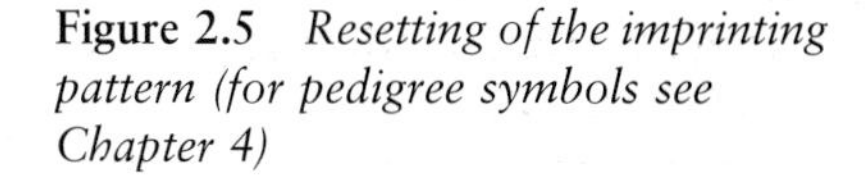

Figure 2.5 *Resetting of the imprinting pattern (for pedigree symbols see Chapter 4)*

The way *XIST* operates is by generating a functional structural RNA, not the usual protein, and its mode of action is to coat the X chromosome which is to be inactivated. As all extra X chromosomes (apart from the active X) are inactivated, there must be a 'counting' or recognition mechanism operating, probably via an autosomal gene.

Complete moles are very rare and account for 1 in 2000 pregnancies in Europe, but are more common in the Far East where they occur with a frequency of 1 in 200 pregnancies. A potentially lethal cancer called a choriocarcinoma may arise, which invades the uterine wall, but this is detectable by monitoring hormone levels.

Not every chromosome is imprinted. Only individual genes or small chromosome regions on particular chromosomes are inactivated.

Mechanisms by which a gene may be imprinted (i.e. switched off or not transcribed) include methylation, configuration of the chromatin structure or the timing of replication.

Evidence for genomic imprinting

Developmental evidence

At fertilization a 23,X or 23,Y haploid sperm fertilizes an egg cell nucleus carrying a 23,X haploid set of chromosomes, which normally produces a 46,XX or 46,XY embryo with its associated placenta and membranes.

If there has been an error at meiosis in the ovum or degradation of the female haploid set, it is possible that a **sperm may enter an empty egg** (which has no chromosomes of its own). If the **haploid set of the sperm doubles**, then we have a structure in which all the chromosomes are **paternally** derived (androgenetic).

This structure, called a **hydatidiform mole**, consists of long strings of **placental** material, comprising many fluid filled cysts. The cysts are derived from chorionic villi, which have no blood vessels and therefore cannot drain fluid. **There is no embryo.**

The paternal genome alone is therefore sufficient for the production of placental structures, but the maternal genome appears to be required for foetal development.

In an **ovarian teratoma** an egg cell bearing a 23,X haploid set of chromosomes doubles to 46,XX, such that all the chromosomes are **maternally** derived (gynogenetic). The egg develops into a round tumour, which if dissected reveals an uncoordinated mass of foetal tissues such as hair, intestines or teeth! There is, however, **no placenta.** This supports the view that the maternal genome is necessary for embryonic development.

Chromosomal evidence: uniparental disomy

Uniparental disomy (UPD) occurs when **both** of a particular pair (or part of a pair) of chromosomes have been inherited from **one parent.** All the remaining pairs of chromosomes have been inherited equally, one from each parent. Some examples of UPD are described below.

Maternal UPD 7

If an embryo has inherited both chromosome 7s from its mother (**maternal UPD 7**), there is a slowing down of the growth of that foetus (interuterine growth retardation, or IUGR). The child is **small** but otherwise unaffected. This phenomenon is not seen if the 7s are both paternal.

Paternal UPD 11

A condition called Beckwith–Wiedemann syndrome may result in a very **large** infant who may have internal organs, such as the liver, greater than normal size. A prominent feature is a large tongue.

There is usually an abnormality of the short arm of chromosome 11. Although Beckwith–Wiedemann can arise via different mechanisms, often **two paternal 11p15 regions** are inherited together with one maternal 11, essentially resulting in **trisomy 11p15**.

It is known that a growth gene (IGF2) is present in this region, which is **active** on the **paternal 11** but **switched off** (imprinted) on the **maternal 11**. If more than one copy of the active paternal gene is inherited, there will be **overgrowth.**

Genomic imprinting is found in X-inactivation in females, the autosomes of somatic cells, and can also be acquired. An example is the retinoblastoma tumour supressor gene (see Chapter 8), whereby the promoter of the remaining normal copy of the *RB1* gene is methylated.

Evidence from imprinting syndromes

Prader–Willi syndrome (PWS) and **Angelman syndrome (AS)** are two phenotypically different syndromes (see Appendix) whose genes map to the same locus: 15q11-13. It is now realized that, although the genes for PWS or AS lie close together, they are coded separately and **differently imprinted**. This can be deduced for the following reasons:

- A deletion of the paternal 15 or maternal UPD results in Prader–Willi syndrome.
- A deletion of the maternal 15 or paternal UPD results in Angelman syndrome (Figure 2.6).

The gene preventing Prader–Willi syndrome is imprinted (switched off) on the maternal 15, and therefore a working paternal 15 is necessary for normal development. The gene preventing Angelman syndrome must be active on the maternal 15.

The long arm of chromosome 15 at 15q11-13 is a well known example of human imprinting. For example, in a **woman's somatic cells** one chromosome 15 would be inherited from her mother (maternally imprinted) and the other would be inherited from her father (paternally imprinted).

In order to have normal children, **her paternal chromosome 15s must be reset as maternal** before being passed on via her germ cells. Similarly, her partner's maternal chromosome 15s in his sperm will have been reset to paternal. That way the embryo will receive a correctly imprinted 15 from each parent.

Genomic imprinting appears to be **specific to mammals**. When a fertilized mammalian egg divides into a ball of cells called a blastocyst, the groups of cells differentiate into two types (lineages). The **inner cell mass will give rise to the embryo, whilst the trophoblast (outer) layer becomes the placenta and membranes**.

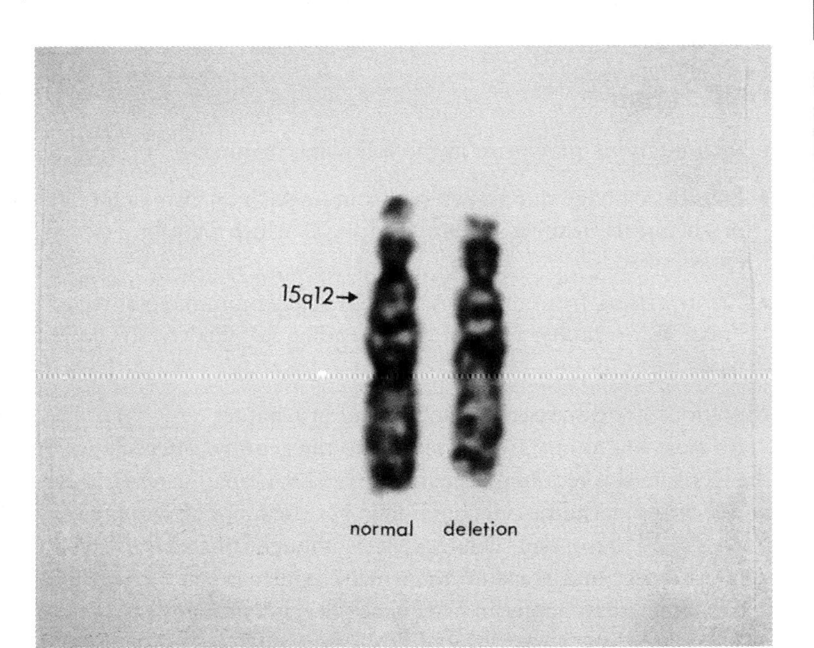

Figure 2.6 *A pair of homologous chromosome 15s, showing the microscopically visible deletion typical of Prader–Willi syndrome and Angelman syndrome*

UPD of a whole chromosome is not always required to produce an effect, and UPD of only one gene or a group of genes can result in a clinical phenotype. For example, a whole extra chromosome 11 is not needed in order to display Beckwith–Wiedemann syndrome; the critical area has been narrowed down to a region at the end of the short arm at 11p15.

Professor Marcus Pembrey has suggested that imprinting may be useful as a **short term adaptation** (two to ten generations) during times of stress or times of plenty.

It was noted that pregnant Dutch women gave birth to smaller infants during the famine of the Second World War. The post-war female infants themselves went on to have small babies, suggesting that this effect may last more than one generation. The opposite is also true. Magellan observed that a sixteenth century Patagonian tribe was renowned for being particularly tall; this no longer remains true in the twenty-first century.

Although it appears that these observations are following acquired inheritance patterns, the explanation may be due to imprinting taking place on particular genes which have been exposed to (for example) a particular maternal diet in the first trimester of pregnancy.

Genes on the short arm of human chromosome 11 (11p15) include an insulin gene known to be capable of imprinting. It has been shown in rats that their insulin sensitivity can be decreased such that two

Why is imprinting required?

Evidence from hydatidiform moles and ovarian teratomas

Genomic imprinting may be needed at a critical stage in cell division when it is important to have only one copy of a specific regulatory gene, such as the divergence of the inner cell mass and the trophoblast layer. Maybe by inactivating (imprinting) one gene or set of genes, a suitable mechanism is created to **avoid confusion and competition between genes.** For example, the maternal genome controls the development of the embryo whilst the paternal genome controls placental development.

Ovarian teratomas are comparatively common in humans, so it has also been proposed that genomic imprinting **protects the female from trophoblastic invasion**, as the active paternal placental genes required for full development are not present. If genomic imprinting was not present, the trophoblast layer would also develop and a lethal choriocarcinoma would result.

Evidence from UPD 7 and Beckwith–Weidemann syndrome

Maternal UPD 7 results in a small foetus, but trisomy 11p15 (which may consist of paternal UPD 11p15 and a maternal chromosome 11) results in a large infant.

It has been suggested that there is **conflict between the mother and her foetus**, such that the paternal genome of that foetus promotes foetal growth (the selfish gene wishing to survive). This may lower the mother's chance of a successful birth, however, so the maternal genome competes by trying to limit foetal growth.

DNA repair

DNA mutations may arise in the following manner:

- Physical damage due to external sources such as ultraviolet light or chemicals, leading to damaged bases which may be fused or cross-linked.
- Errors arising from mistakes in DNA replication, meiotic recombination or faulty DNA repair, leading to incorrectly paired bases.

Mutations are discussed in more detail in Chapter 3.

As most mutations are deleterious to the genome, **surveillance at the DNA level** is required. The fidelity of DNA repair is remarkable: in one year a germline cell containing 3×10^9 bp of DNA may only have 10–20 base pair changes. It is thought that DNA repair systems detect and react to abnormal conformations of the DNA helix, rather than acting in a sequence-dependent manner.

There are three basic types of DNA repair:

generations of their offspring become more prone to diabetes.

The inheritance of these imprinted genes would still be passed on in a Mendelian manner.

Enzymatic photoreactivation (EPR)

This is a **direct** form of repair which **reverses damage** to the DNA. For example, ultraviolet light may give rise to a fusion of two adjacent thymines, known as a **thymine dimer**. An enzyme called photolyase recognizes the distortion in the double helix and, when activated by light, uncouples the dimer. No DNA synthesis is required.

Excision repair

Damaged bases are recognized by enzymes which remove either one damaged base (**base excision repair**) or a length of 27–29 bp surrounding the damaged base (**nucleotide excision repair**, or **NER**).

In NER, the two nicks flanking the base are made by endonucleases, and a multi-subunit complex acts as an exonuclease, which excises the DNA fragment containing the faulty base. The gap is repaired by a DNA polymerase (using the complementary sequence on the other strand) and is closed by a DNA ligase.

Excision repair targets **lesions** such as thymine dimers, alkylated bases, or apurinic sites, but may also be involved in the removal of the normal but mismatched base pairs that are more usually dealt with by MMR (see below).

Mismatch repair (MMR)

The most common source of mismatches arises from DNA replication and, to a lesser extent, during meiotic recombination. Mismatch repair proteins **'proof-read'** the newly synthesized DNA and detect anomalies such as small loops indicative of a mismatch. A nick is made 1–2 kb from the mismatch and the strand is degraded by exonucleases until it passes the mismatch. A DNA polymerase fills in the gaps and inserts the correct base.

The mismatch repair system can recognize the new strand from the parental strand, as the parental strand is methylated at the adenine sites (in the sequence GATC) whereas the newly replicated DNA is not. This type of repair acts as an 'antimutator' pathway.

There may be slippage and mismatching of the microsatellite repeats that are found throughout the human genome, and these will usually be corrected by mismatch repair. At a particular locus these will remain the same length. If mismatch repair is faulty, microsatellite repeats of different lengths are seen. This is the case with hereditary non polyposis colon cancer (HNPCC), discussed in Chapter 8.

Immunogenetics

The immune system

The function of cells in our immune system is to distinguish our own normal cells (self) from both external pathogens such as bacteria and viruses and from internal changes to normal cells such as transformation into cancer cells (non-self).

There are two systems in the body where these fighting cells can be found: the **blood** system and the **lymphatic** system. Examples of lymphatic organs are the bone marrow which manufactures particular undifferentiated stem cells, the thymus (which is found at the base of the neck) and the lymph glands.

The white blood cells (leukocytes) arise from the bone marrow. These are divided into two cell lines:

- The **myeloid** cell line provides a very generalized natural or innate immunity to infection, which is **non-adaptive.**
- The **lymphocytic** lineage provides a more specific immune response, and is part of an **adaptive** system comprising three kinds of lymphocytes, namely T cells, B cells and natural killer cells (or large granular lymphocytes).

Any foreign molecules, including whole bacterial cells, viruses, cellular components or proteins, can be detected by the immune system and are called **antigens.** The response of the adaptive immune system is to produce proteins called **antibodies** which **recognize and bind to the antigen**; these disease-causing molecules are then removed from the system.

Our immune system mainly functions using two of the three types of lymphocytes.

B cells

When a B cell meets an antigen for which it is specific, **clonal selection** occurs, whereby the cell divides into many antibody producing plasma cells. However, a few specialized **memory cells** are produced, which will remember the antigen should it be encountered in the future.

B cells arise in the **bone marrow** and differentiate into plasma cells and memory cells.

Antibodies are the protein product of the B-cell genes; they are called **immunoglobulins**, written Ig. As antigens can appear in the body in vast numbers of different types of molecules, it would be expected that enormous numbers of different antibodies would be needed to fit the antigens.

However, there is not enough DNA to code for the millions of genes necessary to do this. Immunoglobulins therefore have not only common subunits, but also **variable regions** generated by **somatic recombination** of the DNA (i.e. the genes) during the development of the B lymphocytes. To see how this **genetic diversity** is achieved, we must look at antibody structure.

Antibody structure

Immunoglobulins comprise four chains or protein subunits. Because of the differences in weight, the two identical longer chains are called the **heavy chains** (H) and the two shorter chains are **light chains** (L). They are held together with disulphide bonds in a 'Y' shape (Figure 2.7).

The upper region that binds the antibody is called the **Fab** (**f**ragment **a**ntigen **b**inding) fragment; the lower region is the **Fc** (fragment **c**rystallizable) fragment. As expected, the variable region binds the antigen.

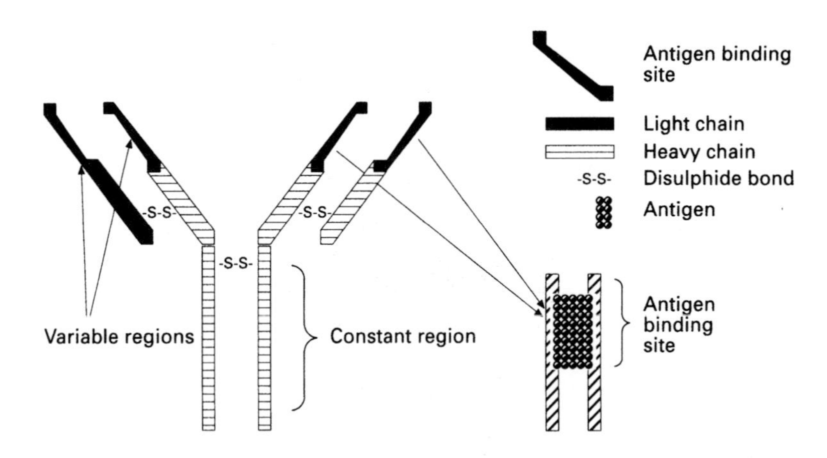

Figure 2.7 *Antibody structure*

Light chains (L)

There are two types of light chains, called kappa (κ) and lambda (λ), coded by genes on chromosomes 2 and 22 respectively. However, although we produce both types of light chain, in any particular immunoglobulin molecule either two κ or two λ chains interact with the heavy chains – never one of each.

Somatic recombination of light chains

If we consider, for example, the production of a kappa light chain, we first have to look at the arrangement of genes in the **germline**. As an embryo, all our cells start with a common arrangement of kappa genes, coding for three different regions of the final chain. As the B lymphocyte differentiates and matures, a **somatic recombination** (rearrangement) occurs which is **unique** in that it involves post-zygotic DNA.

In the unrearranged germline, there are three segments separated by introns: the **variable** light chain region (V_L) comprising around 300 genes, a smaller **joining** segment (J_L) comprising five genes, and a single **constant** region (**C**).

Using enzymes called recombinases, just one Vκ gene is selected, together with just one Jκ gene; these are then joined to the C gene to form a VJC unit which has a fixed antigen specificity in a particular B cell (Figure 2.8).

If any one of the 300 separate Vκ genes can join with any one of five Jκ genes (and one constant), then 1500 different kappa chains are possible.

The rearrangement process in lambda light chains is similar to kappa chain synthesis and gives a similar number of rearrangement possibilities.

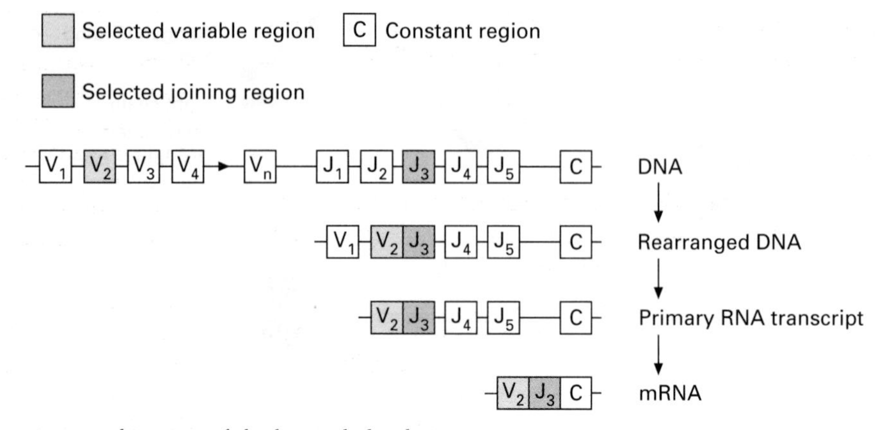

Figure 2.8 *Somatic recombination of the kappa light chain genes*

Heavy chains (H)

These are coded for by genes on chromosome 14. In a similar manner to light chains there is a **variable** region (V) comprising 100–300 genes, an additional **diversity** segment (D) of 12–20 genes, and 4–6 **joining** (J) genes. The **constant** (C) region is separated by a large intron and is more complicated than in the light chains, as there are five C genes in the germ cells.

Somatic recombination of heavy chains

Two rearrangements occur separately: a D gene is joined to a J gene, and then a V gene is brought next to the DJ unit, giving a fixed antigen specificity. A total of at least 4800 different heavy chains are therefore possible.

IgM is the first immunoglobulin made in an immune response by the B cells themselves. It appears on the cell surface, and when a matching antigen is encountered division of that B cell is triggered.

Genetic diversity in B cells

It can now be seen that from a few hundred subunits of the kappa and lambda light chain genes together with the heavy chain genes, as many as 10^{11} combinations are possible. This generates antibodies in a sufficient variety of shapes to bind with different antigens.

T cells

T cells arise in the bone marrow and migrate to the **thymus** to mature. As they do so, they develop membrane receptors (**T-cell receptors**, or **TCRs**) which recognize particular antigens. Unlike the free immunoglobulins of B cells (the humoral response), these receptors remain bound on the surface of the T cell (the cell-mediated response).

The structure of the T-cell receptor includes an α and β protein chain (designated Tiαβ), or a γ and δ protein chain (Tiγδ).

Whilst B cells recognize antigens on their own, the T-cell receptors need the help of other sets of proteins such as those of the **major histocompatibility complex** (MHC).

The MHC region is on the short arm of chromosome 6, comprising a number of genes which are very polymorphic; this means that each gene has up to 50 alternative sequences (alleles) per locus.

Each individual therefore has a near unique combination of these genes which identify you and your cells as **self**. In families, closely related people have more chance of sharing these genes, which are passed down from our parents in 'blocks' or **haplotypes**.

This is why organ transplants are rejected between unrelated individuals, but have more chance of succeeding within a family, as the **tissue type** is more likely to have a closer match of genes. Identical twins, having identical MHC genes, will match exactly. However, **identical twins will have their own non-identical sets of immunoglobulins produced by differing somatic B-cell recombination.**

T-cell diversity

T-cell maturation also involves **rearrangements** of the gene segments, which will result in unique T-cell receptors. γ and δ chains have V and J segments, while α and β chains have V, D and J segments.

If we take the diversity of immunoglobulins (with heavy and light chains) **and** the two types of TCRs (αβ and γδ), the total possible number of rearrangements is enormous. Other mechanisms, such as hypermutation in B cells, raise the number of combinations for antigen recognition to a staggering 10^{18}.

Allelic exclusion

The components synthesized by B and T cells are either derived from the maternal chromosome *or* the paternal chromosome, but not both. One allele is **excluded**.

The chromosomal input into a B cell is from **both** parents – so theoretically there should be a choice of alleles (alternatives) from two parental chromosome 2s, two chromosome 22s and two chromosome 14s respectively.

However, in each B cell only **one locus from each chromosome is expressed**, so that the antibody comprises one heavy chain, together with **either** a kappa **or** lambda light chain.

In a particular T cell, having for example two α and two β chain loci, only **one** α and **one** β locus are expressed, showing the allelic exclusion found in B cells.

The β and δ chain genes are on chromosome 7, and the β and γ chain genes are on chromosome 14.

Mitochondrial genes

In Chapter 1 we introduced mitochondria as organelles found in the cell cytoplasm of eukaryotes. Each cell has several hundred mitochondria, each having an outer smooth membrane and an inner folded membrane, where energy is generated for the body's metabolic processes by oxidative phosphorylation (**OXPHOS**).

The mitochondrial genes are independent of the nuclear genome in that they have their own transcription and translation systems.

OXPHOS involves five multi-enzyme complexes. Thirteen mitochondrial genes code for some of the subunits of four of these complexes: the NADH dehydrogenase complex, the cytochrome b-c1 complex, the cytochrome c oxidase complex and the ATP synthase complex. Together with subunits encoded by the nuclear genome, these components enable molecular oxygen to be released in a usable form from cells.

The reason why mitochondria appear to be more independent than other cell organelles is that they may have been derived from prokaryotic bacteria which became symbiotic (lived together for mutual benefit) with eukaryotic cells early in evolution.

Another way that totipotency can be shown arises from the method commonly known as **cloning**. Originally demonstrated in frogs by Gurdon in the 1960s, it is now possible in mammals. Dolly the cloned sheep was born in 1997, after Ian Wilmut and his veterinary colleagues in Scotland took a mammary gland cell (which had been frozen in liquid nitrogen for 6 years!) derived from the original Dolly. The nucleus was extracted by micromanipulation (using a very narrow pipette). An egg (ovum) from another sheep was taken and the nucleus removed. Dolly's nucleus was then pushed through the egg cytoplasm and given a weak electric shock to simulate fertilization.

The result was an apparently normal sheep, proving that one mammary gland cell can eradicate its mammary gland origin and 'redevelop' (dedifferentiate), thus showing that all the genes must be present, and that the cell nucleus must be totipotent.

The only question remaining was – how old was Dolly? Was she newborn, or was she 6 years old, the age of her original cell? In 1999 it was reported that Dolly was ageing prematurely, which suggests that the age of the donor cell nucleus is not reset.

The other major group of 22 mitochondrial genes code for each tRNA corresponding to a particular amino acid, e.g. $tRNA^{Ser}$ or $tRNA^{Leu}$.

Developmental genetics

The aim of this section is not to describe in detail the development of a foetus; this can be obtained from a good embryology book. We have been looking at how genes are controlled and how they control other genes; this theme is continued here by studying three of the many interacting cascades of genes responsible for certain aspects of developmental pathways: growth factors, signalling genes and homeobox genes.

In the early days of developmental biology, the key question with respect to a newly fertilized embryo was whether each cell (e.g. an intestinal cell or a brain cell) had a predetermined fate, or whether **every** cell had the potential to become **any** of the differentiated body cells.

An early human embryo containing just a few hundred cells can divide into two roughly equal groups, each of which develops into distinct but identical (monozygotic) twins.

All the genes of the dividing embryo are therefore active at this early stage. Embryonic cells are **totipotent**, and have the potential for full growth and differentiation into any cell type.

Three layers of cells arise in the early embryo: the **endoderm** (= inner, giving rise to structures such as the lungs), the **mesoderm** (= middle, giving rise to the blood and genito-urinary system) and the **ectoderm** (= outer, giving rise to the spinal cord, brain and skin).

How do these cells know:

- where they are in the cell mass (**position**);
- which end of structure should grow (**polarity**);
- where in that general structure a pair of limbs or an organ should be situated;
- how many limbs there should be and which way up they should go (**pattern formation**)?

As well as knowing how to differentiate and grow, cells also have to know when to slow down, stop growing or die. A foetal hand starts by looking like a mitten; the fingers are formed because cells die in the spaces in between. This is called programmed cell death (**apoptosis**) and is discussed further in Chapter 8.

Embryonic pattern formation: the limb

Taking the limb as an example of development, each phase can be studied with respect to the particular groups of genes involved at each stage. Although in the past, most developmental biology has been studied through non-human sources such as the fruit fly *Drosophila*, or vertebrates such as the mouse or chicken, it seems likely that animal models will be found to apply to humans too (except for minor differences such as wings or tails!).

Orthologs are similar or homologous genes or DNA sequences between different species, e.g. *SHH* in humans and *hh* in *Drosophila*, or *SRY* in humans and *Sry* in the mouse.

Paralogs are similar or homologous genes or DNA sequences **within a species**, e.g. the four sets of homeobox or *HOX* clusters of human genes, such that A13 is a paralog of D13.

Growth factors

During the initiation of the limb bud, cell division must occur by mitosis. **Growth factors** provide the initial **stimulation** for this to occur and they can also control the cell cycle during the transition from G_0 to G_1. Different kinds of cells have appropriate growth factors; for example, fibroblast growth factors (**FGF**) for fibroblast cells (e.g. connective tissue), epithelial growth factors (**EGF**), nerve growth factors (**NGF**) and platelet-derived growth factor (**PDGF**) which is found in blood vessels and helps in clotting.

Some factors can **suppress** growth – the tumour necrosis factor (TNF) or the transforming growth factor (TGF) are examples.

Each of these growth factors has a membrane-bound cellular **receptor** to which that specific growth factor binds. An example is FGFR3 (fibroblast growth factor receptor 3). Inside the cell (intracytoplasmic) the binding of a growth factor to a specific receptor causes an intracellular cascade of events culminating in chemical signals which can suppress or activate developmental genes.

The hedgehog signalling pathway

Two stages must occur during embryonic pattern formation. The cells must be informed of their place in a three-dimensional system and they must know their spatial orientation. Only then is further information interpreted to form appropriate structures, for example in the specification of limb pattern for an arm or leg.

It is believed that groups of cells have certain boundaries which are characterized by gradients (differences in concentration) of different molecules (sometimes known by the general term **morphogens**). Cells respond to these 'threshold concentrations' which arise from a positional signal from a specialist area of the cell, the **polarizing region** (P).

In the context of the structure of an organism:

- Anterior is the front end – in humans this is the 'head end'.
- Posterior is the hind end – in humans this is the 'tail end'.
- Ventral is the underside – in humans this is the stomach surface.
- Dorsal is the upper surface – in humans this is the back.

Cells near the P region are exposed to high morphogen concentrations and form **posterior** digits (e.g. legs, feet, toes); cells further away become **anterior** digits (arms, hands, fingers). A human gene thought to encode such a long-range signalling molecule across a limb bud is the **sonic hedgehog gene** – *SHH*. This is one of a class of **segmental polarity genes.**

The hegehog signalling pathway is extremely complex, but some elements are understood, mostly through its *Drosophila* counterpart *hh* and its mouse counterpart *Shh*. The sonic hedgehog gene is responsible for the patterning of the ventral neural tube, the notochord, limbs, foregut and lung, controlling growth and differentiation by switching on (inducing) other banks of genes which have more specific functions (Figure 2.9).

The *WNT* (wingless related) gene, for example, is involved in the **ventral** patterning of a limb, and *DPP* (decapentaplegic) in patterning on the **dorsal** side. *HOX* genes have been extensively studied and are discussed below.

Human genes are written in italicized capitals. Their products are written in non-italicized capitals. Genes from non-human species are written in lower case italic script and mouse genes also have an initial italicized capital letter.

Interestingly, expression of *SHH* also induces transcription of a gene called **patched** (*PTCH*) which in a negative feedback pathway acts as a receptor for the SHH product, suppressing *SHH* production and shutting down the hedgehog pathway. In its normal role of growth suppressor, patched has become known as a **'gatekeeper'**: a line of defence against uncontrolled skin cell proliferation resulting in tumours. We will meet this gene again in Chapters 3 and 8.

Homeobox and paired box genes

We have seen how the limb bud is initiated, how the pattern is specified and the correct dorsal and ventral differentiation occurs. How do the arms and legs know **where** to appear along the length of the backbone?

This is the function of the **homeobox** (*HOX*) genes. There are 38 in humans, divided into four groups, A–D, and then numbered (e.g. 1–13), such that a particular gene may be called, for example, *HOXD13*. Originally studied in *Drosophila* larvae where they are responsible for determining the fate of each larval segment in the adult fly (e.g. antennae, wings), in humans they determine at which point along the spinal vertebrae the appropriate limbs appear (anteroposterior position **patterning**), and also ensure the correct numbers of appendages – in matching pairs!

The *HOX* genes are actually transcribed in order (i.e. *HOXA1* to *HOXA13*) in the 5′ to 3′ direction, which corresponds to a linear time sequence from the posterior end of the embryo to the anterior end.

Some functions of an associated group of genes called **paired box** (*PAX*) genes have also been identified. *PAX-3* controls the growth and differentiation of groups of cells which have multiple roles in eye and hair pigmentation and hearing. *PAX-6* is involved in normal eye formation, especially that of the coloured part of the eye, the iris.

Growth to adult size

Once the foetal limbs are fully differentiated, there follows a period of enlargement. Here we have come full circle, as fibroblast growth factors (FGFs) are again required for both shaping and growth.

When considering these stages it becomes apparent that the same sets of molecules can be involved in development at different times and in different places. In this way the numbers of mechanisms required in genetic development are conserved.

Sex differentiation

One of the first questions the proud parents of a newborn baby are asked is 'is it a boy or a girl?'. Most people now know that, of our 46 chromosomes, two are called the sex chromosomes – XX for a girl and XY for a boy. It seems reasonable to assume therefore that the Y chromosome must contain genes specifically for male

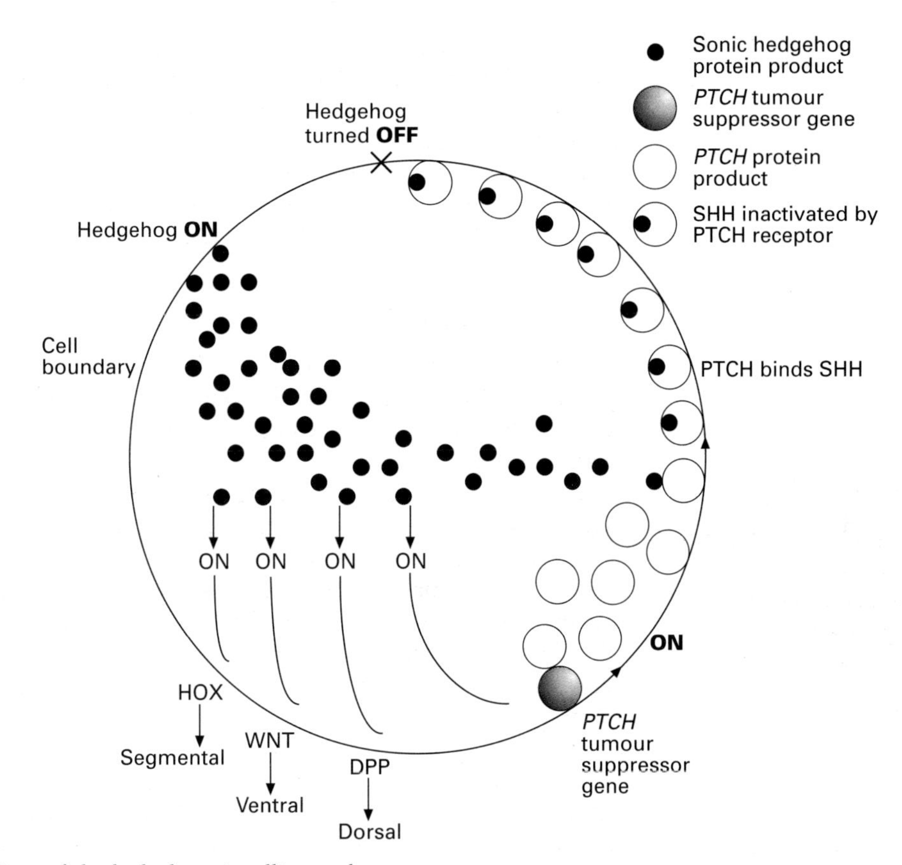

Figure 2.9 *Part of the hedgehog signalling pathway*

development. A male sex determining gene inducing testis formation, lying proximal to the PAR, was found in 1991, called *SRY* (the **s**ex determining **r**egion on the **Y**). Some books refer to this as the *TDF* (**t**estis **d**etermining **f**actor).

> It is possible that early in evolution there were two identical chromosomes similar to the present X chromosome. Over time one of the X chromosomes evolved into the present Y chromosome.
>
> Vestiges of this heritage are still to be found if we look at certain regions of homology on Xp and Yp, where the DNA sequences are so similar that these regions can pair at meiosis. These are the pseudoautosomal regions discussed earlier.

Sexual development in the foetus

At 6 weeks of age, embryos contain identical precursors for both the male and female sexual structures, called respectively the **Wolffian** and **Mullerian ducts**. After 6 weeks the pathways of these **indifferent gonads** begin to diverge, according to the presence or absence of the *SRY* gene on the Y chromosome.

The male pathway

By 10 weeks, if the *SRY* gene is present, the Wolffian ducts will begin to develop into the foetal testes. The testis secretes two hormones, one of which inhibits the development of the Mullerian ducts. The other is the male hormone **testosterone**. Providing the target cells have androgen receptors which will respond to testosterone, differentiation of internal male structures such as the vas deferens, epididymis and seminal vesicles will follow.

At 15–16 weeks some of the testosterone is converted into 5-hydroxytestosterone, which is responsible for the development of the external male sexual structures such as the urethra, scrotum and penis.

The female pathway

In the absence of the *SRY* gene, the Wolffian ducts degenerate, and by 10 weeks the foetal ovary is producing oestradiol, leading to the differentiation of the female structures such as the uterus, Fallopian tubes and upper vagina.

Other genes involved in sexual differentiation

SRY is not the only gene believed to influence sex determination. In Chapter 3 we will look at evidence to suggest that both the correct dosage of a gene on the short arm of the X chromosome and an intact gene called *SOX9* on chromosome 17 are also required for normal male development.

In humans the female is the **homogametic sex** (producing one kind of gamete) and the male the **heterogametic sex** (producing two kinds of gametes). This is not always the case in other species.

In birds, moths and butterflies the sex chromosomes are termed **Z** and **W**. The female is the heterogametic sex (ZW) and the males are homogametic (ZZ).

In *Drosophila*, **sex determination** depends on the **X/autosome ratio**, i.e. the **balance** of female determining genes on the X chromosome(s) and the number of haploid sets of autosomes. The Y plays no part in determining sex. The critical ratio for maleness is 1(X):2 (autosome sets). So, for example:

- a normal female = XX + 2*N* autosomes;
- a normal male = XY + 2*N* autosomes;
- a 'metafemale' = XXX + 2*N* autosomes;
- a 'metamale' = X + 3*N* autosomes.

Summary

When genes were first postulated, the definition 'one gene one polypeptide' or 'one gene, one enzyme' did appear to explain simple linear biochemical pathways. With the concept of gene control came the realization that genes could act in more complex ways, as demonstrated by embryonic development, the immune system, and even by determining which sex we are.

We now know that some genes do not produce proteins – they produce structural RNAs instead, which exert their control by their shape. We cannot even be sure that a gene we receive from one parent will behave in the same way as the allele we receive from the other parent, as demonstrated by imprinting.

Considering the intricate cascades of structural and controlling genes needed for the correct functioning of the human body, it is not surprising that occasionally individuals are born in whom certain genes do not function correctly.

The next chapter looks at some examples of genetic disorders. We see how understanding the mechanisms of genetic malfunction helps us to understand the normal role of genes.

Suggested further reading

Benjamini, E. and Leskowitz, S. (1991). *Immunology: A Short Course*, 2nd Edn. Wiley-Liss.

Carrel, L. and Willard, H.F. (1998). Counting on *XIST*. *Nature Genetics*, **19**, 211–212.
Cohn, M.J. and Tickle, C. (1996). Limbs: a model for pattern formation within the vertebrate body plan. *Trends in Genetics*, **12**(7), 253–257.
Dean, M. (1996). Polarity, proliferation and the *hedgehog* pathway. *Nature Genetics*, **14**, 245–247.
Ellis, N.A. (1998). The war of the sex chromosomes. *Nature Genetics*, **20**, 9–10.
Hammerschmidt, M., Brook, A. and McMahon, A.P. (1997). The world according to *hedgehog*. *Trends in Genetics*, **13**, 14–20.
Ohlsson, R., Hall, K. and Ritzen, M. (1995). *Genomic Imprinting Causes and Consequences*. Cambridge University Press.
Passarge, E. (1995). *Colour Atlas of Genetics*. Thieme.
Wallace, D.C., Brown, M.D. and Lott, M.T. (1997). Mitochondrial genetics, in *Emery and Rimoin's Principles and Practice of Medical Genetics*, 3rd Edn. Churchill Livingstone.
Wolpert, L. (1996). One hundred years of positional information. *Trends in Genetics*, **2**(9), 359–363.

Self-assessment questions

1. Gene A makes a protein that forms the globin of red blood cells. Gene B makes a protein which is a transcription factor that helps to switch on gene C. What is the difference between gene A and gene B? What kinds of genes are they?
2. Why is a primary RNA transcript freshly copied from a DNA template longer than the piece that is eventually translated into protein? Name the process involved.
3. A gene for the enzyme steroid sulphatase is found on the X chromosome very close to the pseudoautosomal region. Females have approximately twice the level of this enzyme as males. As dosage compensation has not occurred, what explanation is possible?
4. Name a major developmental gene and give an example of its function.
5. Name the gene on the Y chromosome responsible for male development.
6. A genetic condition exists called Turner syndrome; the individual has 45 chromosomes including only one X. What sex will they be and why? Do you think that they will be affected in any way?
7. What is somatic recombination and why is it important in antibody production?

Key Concepts and Facts

Types of Genes

- Genes produce either protein subunits or functional RNAs. The proteins may be structural or may control other genes. The RNAs tend to control by acting in a positional (spatial) manner.

Gene Expression

- One strand of the double-stranded DNA comprising a gene is transcribed to produce a primary RNA transcript, which is spliced to form messenger RNA. This is then translated into protein with the help of ribosomes and tRNAs.

Control of Gene Expression

- Although housekeeping genes are always switched on, most genes are switched on or off as required, using various levels of control. This ranges from general control via gene promoters, methylation, gene dosage and chromatin structure, to specific gene control such as using silencers and enhancers.

Genomic Imprinting and UPD

- Some gene expression is dependent on the parent of origin. Because some genes only express on certain chromosomes from one parent, abnormalities may arise due to uniparental disomy, where both chromosomes containing inactive genes are inherited from the same parent.

DNA Repair

- The three basic types of DNA repair are enzymatic photo-reactivation, excision repair and mismatch repair.

Immunogenetics

- The products involved in immune response produced by B and T cells are derived from the maternal or paternal genome by alleleic exclusion. The variety of antibodies (immunoglobulins) produced by B cells could not exist in such large numbers if it were not due to somatic recombination, which uses very few genes which then recombine to produce a diversity not otherwise feasible.

Developmental Genetics

- Early embryonic cells are totipotent. Developmental genes are usually controlled in cascades with appropriate feedback loops. There are genes which initiate cell growth and differentiation, and genes which are responsible for the correct positioning of limbs.

Sex Differentiation

- A normal male karyotype is 46,XY. The male sex determining region is found on the Y chromosome, but it is believed that the correct dosage of certain genes on the X chromosome is also required. A normal female karyotype is 46,XX. Two X chromosomes constitute the correct dosage for the female, as some genes are still active on the inactivated X.

Chapter 3
Mechanisms of disease

Learning objectives

After studying this chapter you should confidently be able to:

List the major types of DNA mutations and give examples.

Outline the concepts 'gain of function' and 'loss of function'.

Describe how gene control may be altered.

Explain the consequences of the failure of resetting of imprinting.

Give examples of immunogenetic and mitochondrial disorders.

Outline the mechanisms by which DNA repair may fail, and give examples.

Describe the results of gene mutations relating to developmental and sex determining pathways.

Genetic diseases may be caused by a number of different mechanisms (see Table 3.1 for a summary) but, at the molecular level, these arise from the heritable alteration of a DNA sequence, known as a mutation.

Mutations

A **mutation** may be defined as any **change in the genetic make-up** of a cell, a population of cells, or an organism. The mutation originates as an error in one cell, and will only have a determinable effect if it is heritable: that is, when that cell divides, the mutation becomes distributed throughout the descendants. It is convenient to identify two broad categories of mutations: molecular (or gene) mutations and gross chromosome mutations.

Many mutations are deleterious; some are extremely important in the aetiology and pathology of human genetic diseases, and it is these that comprise the very basis of this book.

It is important to understand that DNA is not just a large complex molecule – it is active through replication and transcription, and it is chemically reactive. Inevitably, therefore, changes to the DNA occur continually. Mutations, changes that become fixed

Some of the possible **endpoints of mutation** are as follows:

- The accumulation of mutations occurring as errors in DNA replication possibly contributes to cellular senescence. Some mutations are inevitably lethal to the cell in which they occur.
- Mutations which interfere with normal cellular differentiation and proliferation lead to tumours.
- Mutations occurring in the germ cell line are heritable and may lead to genetic disease (or handicap) in the offspring.
- Advantageous mutations may occur and may contribute to evolution through selection. Further discussion of such advantageous mutations is beyond the scope of this book.

Table 3.1 *Mechanisms of change*

Normal process	*Disorder*	*Gene*	*Mechanism*
Gene control and expression			
Transcription	β-Thalassaemia		Base substitutions or frameshift mutations
Dynamic mutations	Fragile X	*FMR-1*	Triplet repeat CGG
	Myotonic dystrophy	*DMPK* (myotonin protein kinase)	Triplet repeat CTG
	Huntington disease	*IT15*	Triplet repeat CAG
	Friedreich ataxia	*X25*	Triplet repeat GAA
Chromatin structure	Facioscapulohumeral dystrophy		Position effect
Dosage	Turner and DiGeorge syndrome		Haploinsufficiency?
X-inactivation	Wiscott-Aldrich syndrome		
Methylation and imprinting	Prader-Willi and Angelman syndrome	*SNRPN?/UBE3A*	Deletion, UPD, imprinting centre mutation
DNA repair	Xeroderma pigmentosum		Mutations in genes involved in excision repair
Immunogenetics	Bruton agammaglobulin-aemia		B-cell defect
	DiGeorge syndrome		T-cell defect
	Severe combined immuno-deficiency disease		B- and T-cell defects
Mitochondrial	MELAS, MERRF, etc.		Mutations in the mitochondrial genome
Development			
Growth factors	Achondroplasia	*FGFR3*	Mutations in fibroblast growth factor receptor genes
	Craniosynostosis (Pfeiffer and Crouzon syndrome)	*FGFR1* and *FGFR2*	
Hedgehog signalling pathway	Holoprosencephaly	(*SHH*) *Hpe3* 7q36	Haploinsufficiency via deletion/positional silencing via telomere
	Basal cell carcinoma	(Patched) *PTCH*	Haploinsufficiency?
Homeobox genes	Synpolydactyly	*HOXD13*	Gain of function
	Waardenburg syndrome	*PAX3* (2q35)	Haploinsufficiency/loss of function
	Aniridia	*PAX6* (11p13)	Loss of function
Sex determination	Sex reversal	*SRY*	Deletion/mutation
	Testicular feminization	Androgen receptor	Deletion
	Congenital adrenal hyperplasia	CAH (*CYP21B*)	Deletion/mutation, gene conversion
	Campomelic dysplasia	*SOX9*	Position effect

and inherited by daughter cells at cell division, originate either as a straightforward error of a normal cellular process such as replication, recombination or mitosis, or as a result of the influence of an external agent, known as a **mutagen**. Most mutagens are agents reacting with DNA, while others may affect the genetic make-up of a cell indirectly by disrupting cellular processes that control the normal behaviour of the genetic material. Familiar examples include ionizing radiation, ultraviolet light, and certain classes of chemicals such as alkylating agents. Many mutagens have been proven also to be **clastogens**, that is they induce breakage of chromosomes, visible at mitosis, and **carcinogens**, inducing tumours (see Chapter 8). The normal cellular responses to these agents, for example DNA repair, are dealt with in Chapter 2.

The link between mutagens, clastogens and carcinogens will become clearer in Chapter 8, where it will be seen that tumorigenesis involves mutation at the molecular and chromosomal level.

At the molecular level, an important category of mutation is **base substitution**, where the replacement of one base by another results in a change in a single coding triplet. The most common outcome of such a mutation is to alter a single amino acid in that protein for which the affected gene codes. This is known as a **missense mutation**. However, a number of other effects are possible. For example, the substitution may create a chain terminator codon (see Chapter 1), a **nonsense mutation**, resulting in a truncated protein.

A base substitution mutation in which one purine (guanine or adenine) replaces the other, or where one pyrimidine (thymine or cytosine) replaces the other, is called a **transition**. Where a purine replaces a pyrimidine, or *vice versa*, the mutation is known as a **transversion**.

Alternatively, a substitution occurring at a critical location may alter the splicing of an exon to intron junction, or one occurring in the promoter region of a gene may alter the level of expression of that gene.

A classic example of a base substitution mutagen is nitrous acid, which is able to deaminate either adenine to hypoxanthine, or cytosine to uracil. Hypoxanthine pairs with cytosine; uracil pairs with adenine.

When adenine is deaminated to hypoxanthine, the NH_2 group at position 6 (see Figure 1.4) is replaced by an oxygen atom.

Other molecular DNA mutations involve the **addition or deletion of bases**. Often the result is a **frameshift mutation**. Loss or gain of one or two bases means that all the triplet codes downstream of the mutation become out of phase, and incorrect amino acids will be added to the protein until a stop codon or a splicing signal is encountered. Only if the addition or deletion comprises three bases, or a multiple of three, can the protein continue to be assembled with just the alteration of a small number of amino acids. The most common European cystic fibrosis mutation, ΔF508, is an example of a three base pair deletion which removes a single amino acid, phenylalanine, from the protein.

Frameshift mutagens include ionizing radiation, one effect of which is to knock out single bases. Certain chemical agents intercalate within the double helix of the DNA, distorting it and making it susceptible to damage. Intercalating agents include some of the fluorescent dyes used in the staining of molecular DNA and chromosomes, for example ethidium bromide and acridine orange.

Molecular mutations can be conveniently illustrated by considering abnormalities of the blood protein haemoglobin. In the normal adult, most of the haemoglobin is haemoglobin A. This molecule is a tetramer constructed from four protein subunits, comprising two α-globin chains, and two β-globin chains, coded for by separate genes on chromosomes 16 and 11 respectively. Very many abnormal types of haemoglobin have been discovered resulting from various point mutations, some of which are benign, others resulting in diseases such as sickle cell anaemia, α-thalassaemia or β-thalassaemia.

The sickle cell mutation is a point mutation changing the sixth amino acid in the β-globin chain from glutamic acid to valine as a result of a transversion mutation changing the codon from GAG to

In certain African populations, the sickle cell mutation has reached a high level as a result of the selective advantage of the heterozygous state, or sickle cell trait, which confers resistance to malaria. In the homozygous state, the mutation results in a severe and life-threatening chronic haemolytic anaemia. In some Mediterranean populations, thalassaemia mutations are common, probably also related to malaria resistance, and again mutation in the homozygous state is either lethal or severely life-threatening. In Cyprus, for example, one person in six is heterozygous for a β-thalassaemia mutation.

Colchicine, an alkaloid chemical extracted from the autumn crocus, inhibits production of the mitotic spindle fibres. It is used routinely in preparing cultured cells for chromosome analysis (Chapter 5), since it effectively blocks the cell cycle at metaphase.

Mutagenic and clastogenic agents are commonly used in cancer therapy, where damage to the nuclear DNA of the malignant cells is an effective way of reducing the uncontrolled proliferation of those cells. Radiotherapy with X-rays or radioisotopes is frequently used, while the drugs employed include alkylating agents such as cyclophosphamide, mitotic spindle inhibitors such as vincristine, and DNA synthesis inhibitors, for example hydroxyurea. Patients treated with some types of mutagenic agents have an increased risk of developing a different, unrelated, treatment-induced malignancy.

GTG. The single amino acid difference is sufficient to alter the properties of the haemoglobin molecule.

Some of the common Mediterranean β-thalassaemia mutations involve base substitution at the splice site between the first exon and the first intron, resulting in abnormal processing of the messenger RNA.

The most common β-globin mutation in Sardinia is a transition at codon 39 changing CAG (glutamine) to TAG, which is one of the stop codons. In the homozygote, β-globin is therefore absent.

An example of a frameshift occurring in some Mediterranean β-thalassaemia patients is the loss of A from the codon 6 triplet GAG.

Mutation at the level of the chromosome generates aneuploidy, polyploidy and all of the various types of structural chromosome abnormality discussed in Chapter 5. The external influences leading to chromosome gain are poorly understood, although various substances are known to interfere with normal spindle fibre formation, and thus affect normal segregation of the chromosomes at cell division.

Structural chromosome changes are induced by ionizing radiation, by ultraviolet light, and by chemical agents, which either damage bases or cross-link them. Chromosome abnormalities, for example translocations, are generated by incorrect replication, or incorrect repair, in the vicinity of two or more sites of damage, or lesions.

Triplet repeats/dynamic mutations

The human genome has many types of repeat sequences or 'repeats'. One type of repeats comprises repetitive runs of three DNA bases called **trinucleotide** or **triplet repeats**. Triplet repeat disorders usually show a dominant pattern of inheritance (see also Chapter 4). Normal individuals usually have low numbers of repeats and they have no clinical effect.

As the numbers of repeats rise above a critical level, a pathogenic effect may occur which results in phenotypic expression of a disease. Children who inherit larger numbers of triplet repeats are more likely to show a phenotypic effect.

It has been observed that in most of these disorders the number of repeats increases with each generation, such that the likelihood of disease expression (or severity of disease phenotype) increases down the generations; this is called **anticipation**.

Because the triplet repeats change from generation to generation, they are also known as **dynamic mutations**.

The changes in numbers may arise from mispairing of the two DNA strands as there are so many identical short sequences. When replication occurs the polymerase enzyme copies a longer length of repeats due to the **slippage** and **mispairing** of the strands.

If unequal crossover occurs between the homologues at meiosis, or unequal exchange occurs between chromatids at mitosis (sister

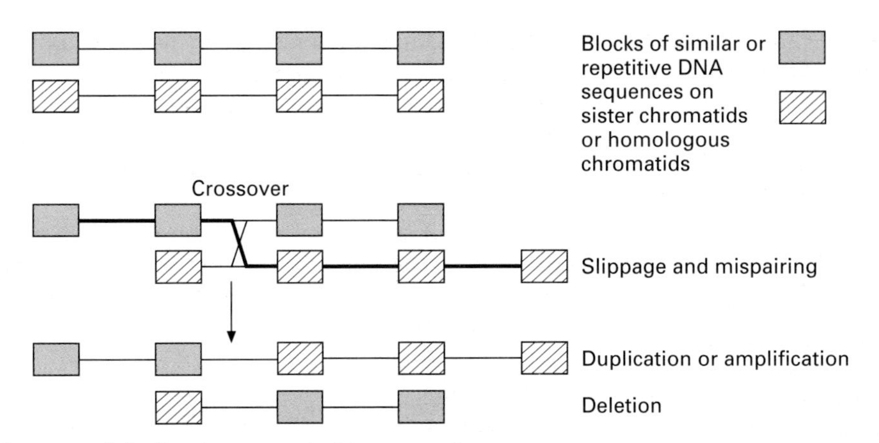

Figure 3.1 *Deletions and duplications generated by unequal crossover*

chromatid exchange), there will be a duplication (or increase in repeats) on one homologue or chromatid, and a deletion (decrease) on the other (Figure 3.1).

The number of diseases caused by triplet repeat **expansion** is currently approaching 20. Huntington disease (chromosome 4), and Machado–Joseph disease (chromosome 14) are caused by an expansion of the triplet repeat **CAG**. Fragile X syndrome is due to an expansion of a **CGG** repeat, and myotonic dystrophy has a **CTG** repeat. The location of triplet repeats (with respect to the gene structure) appears to determine the maximum number of repeats found in an affected person and the disease mechanism.

An exception to the usual pattern of dominant inheritance is Friedreich ataxia, with the novel repeat **GAA**. The autosomal recessive inheritance pattern suggests a different disease mechanism for the GAA repeat, as two copies of the expansion are required. The fragile X syndrome also has an unusual inheritance pattern which is discussed in Chapters 4 and 7.

If triplet repeats are located in the **coding region of a gene**, they are transcribed along with the unique sequences, and this may limit the expansion size. The maximum number reported so far in **Huntington disease** is 121 repeats.

If triplet repeats are in **non-coding regions**, they may show huge expansions up to 2000 repeats, and also very large size changes within a single generation. The expanded repeats may either have **no effect on transcription**, or they may **switch off** the gene by, for example, **methylation**.

In **myotonic dystrophy**, the triplet repeat CTG is in the **3′ untranslated region** of the protein kinase gene on chromosome 19. Affected patients may have very large expansions up to 2000 repeats. The disease mechanism is as yet undetermined. Its dominant inheritance is difficult to understand, as the mutated RNA is not translated into protein. It is possible that the RNA containing the abnormal expansion exerts a dominant negative effect on normal RNA metabolism. It is also possible that the expansion affects

Alteration of gene control

Positional

The normal position or location of a gene or part of a gene may be altered directly or indirectly by changes in **chromatin structure**, or following a **large structural alteration** in the chromosomes such as a translocation between two chromosomes, an insertion, or an inversion of one chromosome.

For example, some female carriers were found to be affected with the X-linked condition Duchenne muscular dystrophy (usually only males are affected) because of a translocation between one X and an autosome. One breakpoint was always on the short arm of the X chromosome at Xp21. This was the location of the DMD gene; it additional as yet uncharacterized genes in this chromosome region.

In the **fragile X syndrome**, the triplet repeat CGG is found in the **5′ untranslated region**. Affected patients have between 200 and 1000 repeats, which affect transcription as the FMR-1 gene is **silenced by methylation**. It appears that the promoter becomes methylated, but the exact mechanism is unknown. It is suspected that both timing of replication and **chromatin structure** are **altered** (hence the **fragile site** on the X chromosome). Under these conditions other genes nearby may also be switched off.

Some genes code for **functional RNAs**, which may be transcribed but not translated. These are known to be important in such genes as *XIST* on the X chromosome and *H19* on chromosome 11. The transcribed sequences of RNA (transcripts) are necessary for regional transcriptional control, and **only work within a particular chromosomal configuration** (for example, relating to **chromatin structure**).

was being disrupted in some way by the translocation such that the normal protein dystrophin was no longer being made.

Therefore it appears that, even though an individual carries an apparently balanced translocation, **small positional changes** at the molecular level may have unexpected effects when one gene is moved next to a new neighbour.

Other small conformational or **positional changes** may involve separating a gene from its promoter or enhancer, which might turn it off, or separating a gene from a silencer, which may turn it on. This can lead to:

- **Gene enhancing**, in which a gene may continue to function after its usual time-span or in the wrong tissues, or is otherwise turned on inappropriately. This can be thought of as a **gain of function.** This is a common mechanism in developmental abnormalities, and is the cause of some cancers (see Chapter 8).
- **Gene silencing**, in which a gene is turned off at an inappropriate time or in the wrong tissues. This can be thought of as a **loss of function.**

Dosage

If the ideal state in the nucleus of each cell is the diploid karyotype comprising pairs of chromosomes, then the genes are present in the correct dosage. Problems can arise when there are either more or fewer genes present.

Syndromes due to the presence of abnormal gene dosage

These may occur when an extra complete chromosome is present, for example in trisomies such as Down syndrome (trisomy 21), Edwards syndrome (trisomy 18) and Patau syndrome (trisomy 13).

An extreme example is seen in the condition known as triploidy, when a whole extra set of chromosomes is present, resulting in 69 chromosomes.

Haploinsufficiency occurs when, despite **the presence of a 50% level of gene product** from one normal copy of a gene, this **is not enough to prevent phenotypic expression.** Examples are DiGeorge syndrome and Turner syndrome.

DiGeorge syndrome

DiGeorge syndrome is associated with a deletion on chromosome 22 of one or more candidate genes. As the phenotype includes several unrelated effects (see the section on developmental genes), it is thought that haploinsufficiency of one or more genes is involved, which fail to act at a critical early stage such that a cascade of gene expression is disrupted. This results in multiple problems further on in development. We will meet DiGeorge syndrome again later in this chapter.

Turner syndrome

Girls with Turner syndrome (45,X) have an abnormal phenotype (see Appendix).

Dosage compensation usually ensures that one of the two female X chromosomes is inactivated such that the genes expressing on the remaining active X produce an equivalent amount of gene product to the single X of the male.

The example provided by Turner syndrome implies that females **do** need either extra structural or controlling gene products normally provided by the missing X. It would appear that a 50% level of certain gene products is not enough, and Turner syndrome should also be considered as a **type of haploinsufficiency.**

Gene silencing may occur if an otherwise active gene is moved to reside in an area that is usually inactive, such as the centromeric heterochromatin or the telomeres of a chromosome. The autosomal dominant disorder facioscapulo-humeral dystrophy (FSHD) at 4q35 may be caused by such a position effect.

The FSHD gene maps close to the 4q telomere and is only separated from it by a large number of 3.3 kb repetitive sequences. Affected individuals have deletions of some of these repeats, bringing the gene closer to the 4q telomere and thus silencing the gene.

X-inactivation

As long as X inactivation is random and haploinsufficiency is not involved, a carrier female of an X-linked recessive disease should be unaffected (see Chapter 4 on Inheritance patterns). Most of the time this is the case.

Skewed X-inactivation patterns

In the presence of an abnormal X, it is possible to get skewing (biasing) of the X-inactivation pattern so the abnormal X is preferentially inactivated. If a female has a deletion of the X chromosome, it is likely that this abnormal chromosome will be inactivated rather than the normal X, thus leaving a comparatively normal genome.

However, circumstances may arise where, due to an X;autosome translocation, the autosomal genes now attached to part of the X chromosome need to remain active. The genetic balance of the karyotype is also important, so the normal X may be preferentially inactivated. However, as we have seen with females affected with Duchenne muscular dystrophy, this may lead to positional anomalies and hence disease expression.

Skewed X-inactivation can also be observed in the Wiskott–Aldrich syndrome (WAS). Symptoms include immune defects, eczema and thrombocytopenia (abnormally long clotting time due to fewer platelets). Some patients develop malignancies; some die of infections or bleeding. As the fault is on the X chromosome (Xp11), males are affected but females are carriers.

Using methylation analysis with DNA probes such as M27β, it was shown that there was a highly skewed X-inactivation pattern in the lymphocytes of female WAS carriers (such that the abnormal X was inactivated). So biased is the X-inactivation in the white blood cells that it can be used as a method of carrier detection using methylation-sensitive restriction enzymes such as Hpa II.

A **candidate gene** is any gene which can be considered as a **possible locus** for a particular disorder.

In a normal female it is known that the tip of the short arm of the inactivated X (Xp22.3) is **not inactivated** (i.e. the PAR), so that two copies of the genes from this region at least are required for normal female development.

Female carriers of the full mutation for the fragile X syndrome may either be normal or mentally retarded. Studies have indicated that the normal females may preferentially inactivate the fragile (abnormal) X, whereas mentally retarded females may preferentially inactivate the normal X.

X-linked agammaglobulinaemia shows non-random X-inactivation in the B lymphocytes of obligate carrier females.

X-linked SCID (severe combined immune deficiency) shows non-random X-inactivation in the T and B lymphocytes and the natural killer cells of obligate carrier females.

Methylation and imprinting

Methylation is often employed in the control of gene expression. One example in which a pathogenic effect may arise due to inappropriate methylation is following the amplification of CGG repeats in the fragile X syndrome, when methylation inactivates the adjacent *FMR-1* gene.

Methylation may also be one of the mechanisms involved in normal imprinting (see Chapter 2). Chromosome 15 is known to be imprinted in humans at 15q11-13. DNA studies have shown that there exists an **imprinting centre (IC)** in this region, which possibly causes a controlled methylation of the genes in the surrounding area.

A mutation in the imprinting centre will prevent the correct resetting of the parental imprint in the germ cells. For example, if a maternally inherited IC mutation is present in the father, that maternal 15 will not reset to 'paternal' and, if inherited by his children, Prader–Willi syndrome will result, as there is no functioning paternal locus.

DNA repair defects in human disease

Human disorders arising from defects of DNA repair often display common characteristics such as increased rates of cancers and abnormalities of the immune response. Mutations of excision and mismatch repair genes result in uncorrected base changes which lead to an increased background mutation rate in other genes, predisposing cells to malignant changes (see Chapter 8).

Xeroderma pigmentosum (XP)

Xeroderma pigmentosum is a very rare (1 per 250 000) autosomal recessive disorder arising in childhood. It is characterized by extreme photosensitivity to ultraviolet (UV) light affecting exposed areas such as the skin and eyes, resulting in basal cell carcinomas and melanomas, together with abnormal pigmentation and neurological abnormalities.

There are eight XP groups, represented by eight different gene loci which code for the different enzyme subunits in the excision complex. Mutations in the genes *XPA–XPG* result in **defects in the initial steps of nucleotide excision repair** of the UV-induced damage, while mutations in *XPV* (variant) lead to abnormalities of post-replication repair. Cells from different XP groups can **complement** one another, restoring levels of DNA repair to normal.

Normal cells synthesize DNA in the S phase of the cell cycle, and also exhibit **unscheduled DNA synthesis** (UDS) when undergoing DNA repair. In the laboratory XP can be demonstrated by exposing **XP cells** to UV light; they **do not undergo UDS**. Although an increase in sister chromatid exchanges and chromatid aberrations can be

Table 3.2 *Disorders of DNA repair and replication*

Disorder	*Frequency*	*Clinical features*	*Mechanism*	*Method of analysis*
Ataxia telangectasia (AT)	1/100 000–1/300 000	Sensitivity to ionizing radiation, hence cancer, immune problems, diabetes, ataxia, redness due to dilation of blood vessels	Deficiency in ATM kinase involved in cell cycle and cell signalling	Expose chromosomes to radiation in G_2; look for an increase in chromatid damage
Bloom syndrome (BS)	100 living cases	Tumours and leukaemias	Defective helicase in DNA replication	Spontaneous increase in SCEs
Fanconi anaemia (FA)	1/50 000–1/100 000	Sensitive to alkylating (cross-linking) agents. Absent radius and thumbs, pancytopaenia. Acute myeloid leukaemia	Molecular basis unknown	Spontaneous increase in chromatid breaks and exchanges; expose chromosomes to alkylating agents

shown, there is no definitive chromosomal test for XP, although DNA sequencing may reveal a mutation in a particular XP gene.

Other disorders of DNA repair and replication

There are a number of other recessive syndromes in which there is a defect in some aspect of DNA repair or DNA replication. In some of these conditions the exact genetic mechanisms are not fully understood. Although they are clinically distinct syndromes there are certain features in common; features that might be expected of syndromes in which an abnormally high level of mutation occurs.

Bloom syndrome, ataxia telangectasia and Fanconi anaemia are often grouped as the 'chromosome breakage syndromes' as they show an elevated level of chromosome damage in culture. They also share a predisposition to tumours and growth retardation (Table 3.2). Cockayne syndrome and Werner syndrome are conditions associated with growth retardation and premature ageing.

Immunogenetic diseases

Disorders of immunodeficiency may affect the B cells (and therefore antibodies), the T cells (and therefore cellular and cytokine response) or both.

An example of B-cell disease

X-linked infantile agammaglobulinaemia (Bruton agammaglobulinaemia) arises from a faulty gene or genes on the X chromosome

Sister chromatid exchanges (SCEs) are exchanges occurring during synthesis between all four strands of the DNA, i.e. the two parental strands and the two newly synthesized, or nascent, strands. This is thought to represent a mechanism by which DNA in the S phase is able to replicate around sites of unrepaired damage with a low risk of introducing mutations into the daughter chromosomes.

Bromodeoxyuridine, or 5′BrdU, can be incorporated into replicating DNA. The 5′BrdU molecule is identical to the nucleoside thymidine except that a bromine atom replaces the CH_3 methyl group (see thymine in Figure 1.4). It is said to be a **thymidine analogue**. The presence of 5′BrdU affects the stability of the DNA and thus alters the staining properties of the chromosome at metaphase. In a culture where two cell cycles take place in the presence of 5′BrdU, only one of the two chromatids of each metaphase chromosome contains a grandparental DNA strand without 5′BrdU. Exploiting this phenomenon, it is possible to stain the two chromatids of a metaphase chromosome differently and see the points at which the pale and darkly staining chromatids exchange.

such that virtually no immunoglobulins (antibodies) are produced. The fault prevents pre-B cells developing into mature B cells, and is a failure of the variable (V) gene rearrangement mechanism. Absence of an appropriate enzyme results in failure of the V_H (heavy chain) genes to join to the D and J genes.

Male infants of 6 months and over exhibit recurrent bacterial infections, and need injections of IgG.

An example of T-cell disease

DiGeorge syndrome is an example of a T-cell deficiency disease. The absence of thymus and parathyroids results in hypocalcaemia (low calcium) and patients have virtually no T cells.

Babies are prone to viral, protozoal, fungal and bacterial infections. As there are no helper T cells, no cytokines (actually lymphokines) can be produced to activate the B cells to produce antibodies in the blood.

In about half the patients with autosomal recessive SCID there is a deficiency of the enzyme adenosine deaminase (**ADA**). Because the gene location is known, the cloned gene has been artificially introduced into a child and is the first example of **successful gene therapy**.

An example of B-cell and T-cell deficiency

Severe combined immunodeficiency disease (SCID) results from the failure of **stem cells** to differentiate into B cells and T cells. There is more than one chromosome location for SCID; one type is on the X chromosome and another is on chromosome 8.

As there are no mature B and T cells, infants are prone to microbial infections. The treatment is usually bone marrow transplantation.

Mitochondrial mutations

Mitochondrial disorders may arise due to faults with mitochondrial structure, altered numbers of mitochondrial genes or impairment of mitochondrial gene function due to DNA mutation.

The organs most affected by mitochondrial diseases tend to be muscles (including the heart), kidneys and liver, as they are the most dependent on the energy source derived from the mitochondrial gene processes. The eyes, ears, neurological system and endocrine system may also be affected (see Table 3.3).

As cloning involves the removal of a nucleus from an egg and the insertion of a somatic cell nucleus into empty cytoplasm, there arises the possibility of gene therapy for mitochondrial disorders.

If a woman had abnormal mitochondria, she could theoretically have one of her healthy cell nuclei transplanted into donor cytoplasm, which would contain normal mitochondria.

As there are so many mitochondria, if a mutation arises in one mitochondrion it may replicate and exist in the same cell together with normal mitochondria. This is known as **heteroplasmy**. After many cell divisions there may exist different ratios of abnormal to normal mitochondria.

As well as mutations in mitochondrial DNA, tRNA and nuclear DNA relating to these conditions, random somatic mutations may gradually accumulate over time. Mitochondrial DNA may mutate up to 17 times faster than nuclear DNA, so over a lifetime these

Table 3.3 *Mitochondrial DNA disorders*

Disorder (abbreviation)	*Full name*	*Clinical symptoms*
MELAS	Mitochondrial encephalomyopathy, lactic acidosis and stroke-like episodes	See left
MERRF	Myoclonic epilepsy and mitochondrial myopathy (with ragged red fibres)	Jerking, hearing loss, ataxia, renal abnormalities, diabetes, cardio-myopathy, dementia – maternal lineage
LHON	Leber's hereditary optic neuropathy	Central vision loss (blindness), sometimes neuro-logical effects on movement. Related through maternal lineage though more males are affected
NARP	Neurogenic muscle weakness, ataxia and retinitis pigmentosa	See left
Leigh		Ataxia, hypotonia, developmental delay, regression, optic atrophy, respiratory abnormalities. Onset 1.5 years.

somatic mutations can reach a threshold at which normal metabolic function may be impaired.

It is believed that mitochondrial mutations may be implicated in diseases of old age such as Parkinson disease and Alzheimer disease.

Developmental changes

Developmental genes tend to act in cascades. When the disruption of one gene causes a multitude of apparently unrelated effects, it is often because that gene is involved in the early development of certain embryonic structures derived from a common origin.

A gene which when expressed results in multiple phenotypic features is said to exhibit **pleiotropy.**

Growth factor receptors

Growth factors are required both at the start of embryonic development and towards the end of that process, when finely controlled proliferation and differentiation occur.

Two examples of pleiotropy are the autosomal dominant Marfan syndrome and DiGeorge syndrome.

- **Marfan syndrome** (chromosome 15): Because the faulty fibrillin gene alters the elasticity of the connective tissue, several apparently unrelated symptoms are seen. These include long limbs and fingers, a defective major blood vessel of the heart (the aorta), and dislocation of the lens of the eye. Fibrillin is found in all of these structures.
- **DiGeorge syndrome** (chromosome 22): Migration of neural crest cells cannot be sustained early in embryonic development, which affects structures called the third and fourth pharyngeal pouches. Structures derived from these are affected in turn, such as the thymus, parathyroids, heart and face.

There are four fibroblast growth factor receptor genes:

- *FGFR 1* is on chromosome 8;
- *FGFR 2* is on chromosome 10;
- *FGFR 3* is on chromosome 4;
- *FGFR 4* is on chromosome 5.

Mutations in *FGFR 1* have been found in Pfeiffer syndrome. Phenotypic features include broad thumbs, deformed great toes, short head, wideset eyes (hypertelorism) and generally a normal IQ.

Mutations in *FGFR 2* have been found in Crouzon syndrome, where the patient has abnormal facial features including premature skull fusion, prominent eyes, beaky nose and poor bite due to

In fruit fly larva, loss of function equates with fewer segments, whereas gain of function equates with more segments.

Achondroplasia is an autosomal dominant skeletal dysplasia resulting in dwarfism due to shortening of the limb bones (the 'long' bones) where the gene is expressed. Mutations resulting in achondroplasia have been found in the fibroblast growth factor receptor gene *FGFR 3*.

There is another group of autosomal dominant syndromes which result from premature fusion of the skull bones. The name for this abnormal skull growth is **craniosynostosis.**

The fusion results in excessive growth elsewhere in the head region, with the result that the head shape is often abnormal. The primitive embryonic cells do not differentiate appropriately into osteoblasts, which make bone.

Many of the craniosynostoses also involve limb abnormalities, which implies that face and limb development share at least one common pathway.

Because of the fusion of the skull bones, the extra digits and syndactyly (fusion of digits), these mutations are considered to result in a **gain of function.** This contrasts with nonsense mutations resulting in haploinsufficiency and a **loss of function.**

Hedgehog signalling pathway

Sonic hedgehog is one of the segmental polarity genes needed for normal cell growth and development of the notochord (spinal cord/backbone), foregut and limbs. It is also important in ventral midline differentiation.

Mutations have now been found in one sonic hedgehog gene, *HPE3,* at 7q36, which results in **holoprosencephaly**. This is expressed as abnormal facial features such as a single eye or absent nose, the fault arising from aberrant movement of cells around a vertical central section of the head. The mutation mechanism is thought to be haploinsufficiency via deletions or positional silencing (7q36 is near the telomere).

Mutations have also been found in the patched gene (*PTCH*), which controls some aspects of cell growth via the hedgehog signalling pathway. As PTCH normally shuts down the hedgehog signalling pathway it follows that a mutation in *PTCH* would lead to uncontrolled cell growth – i.e. cancer.

Basal cell naevus carcinoma syndrome (also known as NBCC, or naevoid basal cell carcinoma) exhibits mutations in the *PTCH* gene. Besides basal cell carcinomas there may also be facial and cranial alteration and overgrowth. The mechanism is probably haploinsufficiency.

In Chapter 8 we will explore the role of *PTCH* as a 'gatekeeper', keeping cancer at bay.

Homeobox genes

These organize the correct anterior and posterior spatial expression of body architecture such as limb position.

Mutations in the *HOX* or related *PAX* genes would be expected to affect specific areas of the embryo which would be reflected in the clinical symptoms produced.

Mutations in *HOXD13* result in synpolydactyly such that there is fusion of the third and fourth digits of the hands or feet, with an extra digit in between. This would be a gain of function.

Mutations in *PAX6* result in the human eye condition aniridia (loss of the iris) and therefore a loss of function. Mutations in *PAX3* result in Waardenburg syndrome type 1. This pleiotropic gene must be important in neural tube development in the brain area, as affected individuals display a variety of symptoms such as deafness, mixed colours in the iris of the eye, and a white forelock. This may result from haploinsufficiency and is therefore a loss of function.

Sex determination

Sex reversal

One particular type of sex reversal is due to unequal crossover between the X and Y chromosome at male meiosis, such that the inheritance of a translocated sex chromosome appears to result in either females with a male karyotype (46,XY) or males with a female karyotype (46,XX).

Although a **phenotypically male** child may appear to have a 46,XX karyotype at the microscopic level, at a molecular level the presence of the ***SRY* gene** inherited on the **X chromosome** results in **male development**. In the reciprocal arrangement, **phenotypic females** appear to have a 46,XY karyotype microscopically, but the ***SRY*** is **not present** on the inherited **Y chromosome,** so **female development occurs.**

De novo mutations in the *SRY* gene can also result in 46,XY females.

Evidence of other genes influencing sex determination

There are cases of phenotypic females and abnormal males who have intact *SRY* genes, but **duplication** of the **Xp21-22.3.** One name for this gene is ***DSS*** (dosage sensitive sex reversal). Thus **gene dosage** must also play a part in sex determination.

It is also postulated that autosomal genes may also be required in the pathway to normal sexual development. Campomelic dysplasia is a skeletal malformation syndrome in which 75% of patients with a **male karyotype** show **sex reversal.**

Mutations have been found in a gene called ***SOX9*** (Sry HMG box) which is found on chromosome 17.

SOX9 shares some homology with the *SRY* gene (a motif called the HMG box) and probably interacts with *SRY* in the developmental cascade.

There are instances of individuals affected with campomelic dysplasia displaying translocations with breakpoints over 50 kb away from *SOX9*; this suggests that **position effect** was the most likely mechanism in these patients.

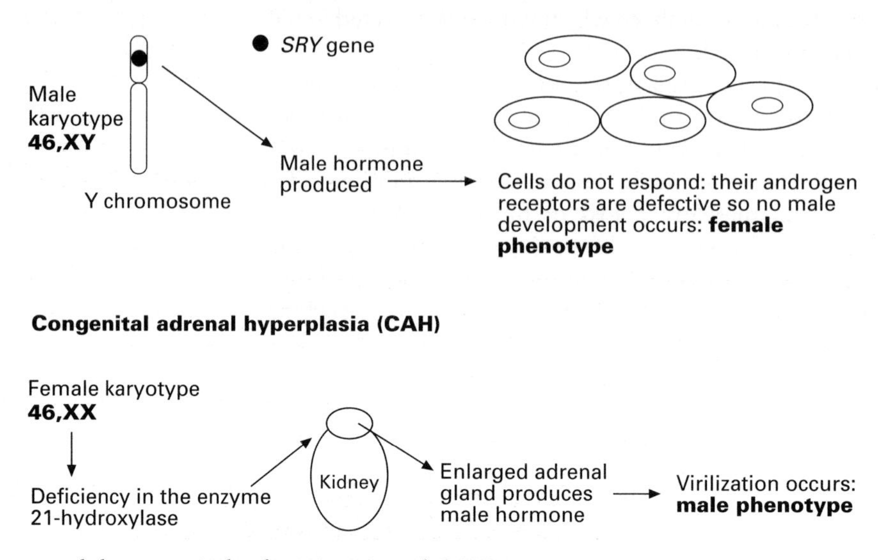

Figure 3.2 *Sex reversal due to testicular feminization and CAH*

Sex reversal due to other gene mutations

Testicular feminization

Males may also display a female phenotype in the presence of an intact *SRY* due to the phenomenon of testicular feminization (Figure 3.2).

Even when male genes produce the correct male hormones, normal male development depends on the target cells responding correctly.

Testicular feminization occurs when the target cells of the male hormone testosterone have a deletion of the androgen receptor. Male sexual differentiation does not occur and the individual appears phenotypically female.

Congenital adrenal hyperplasia (CAH)

Steroid 21-hydroxylase deficiency is just one example of a masculinizing CAH. Both boys and girls may inherit the gene, but in the female the deficiency of the hormone 21-hydroxylase results in overproduction of male hormone from the adrenal glands, with differing degrees of male development (Figure 3.2).

The mechanism by which this is believed to arise derives from the fact that the working copy of the CAH gene (called *CYP21B*) is closely mimicked by a nearby pseudogene (a non-functional gene with a similar DNA sequence) called *CYP21A*.

An unequal crossover (or unequal sister chromatid exchange) can result in a product involving a deletion of the functional *21B* gene – or sometimes a non-functioning *21A/21B* gene.

CAH also illustrates the mechanism of **gene conversion.** Point mutations in the pseudogene *21A* may be copied into the normal *21B* gene. This non-functional inserted DNA replaces a short stretch of *21B*, and gene conversion occurs.

Another example of gene conversion is adult polycystic kidney disease type 1 (APKD1). Mutations in the gene had always been detected with disappointingly low frequency, until it was pointed out that there are three copies of certain sequences near the real gene which resemble it so closely (70% of the sequence is identical) that the potential for unequal crossover or abnormal copying could occur.

Summary

In order to understand the likely effect of gene mutation, we have to understand both the normal function of that gene, and whether the type of alteration is serious enough to impair that function.

A truncated protein may be less damaging than an altered protein, especially compared to the pleiotropic effect of a mutation in a single developmental gene (resulting in multiple malformations due to its importance in a particular cascade of pathways).

Genes which are expressed in the wrong tissue or at the wrong time may lead to deleterious effects from gain of function, while a deleted gene may lead to loss of function.

The phenomenon of triplet repeats and gene conversion from pseudogenes may lead to the discovery of other types of repeat sequences as disease, causing mechanisms; even introns may not be seen as the harmless 'junk DNA' we have supposed them to be.

Suggested further reading

Bardoni, B., Zanaria, E., Guioli, S., Floridia, G. *et al.* (1994). A dosage sensitive locus at Xp21 is involved in male to female sex reversal. *Nature Genetics*, **7**, 497–501.

Belloni, E., Muenke, M., Roessler, E., Traverso, G. *et al.* (1996). Identification of Sonic Hedgehog as a candidate gene responsible for holoprosencephaly. *Nature Genetics*, **14**, 353–355.

Foster, J.W., Dominguez-Steglich, M.A., Guioli, S., Kwok, C. *et al.* (1994). Campomelic dysplasia and autosomal sex reversal caused by mutations in an *SRY*-related gene. *Nature*, **372**, 525–529.

Goff, D., and Tabin, C.J. (1996). *Hox* mutations *au naturel. Nature Genetics*, **13**, 256–258.

Koopman, P., Gubbay, J., Vivian, N., Goodfellow, P. *et al.* (1991). Male development of chromosomally female mice transgenic for *sry*. *Nature*, **351**, 117–121.

Marshall Graves, J.A. (1997). Two uses for old *SOX*. *Nature Genetics*, **16**, 114–115.

Roessler, E.T., Belloni, E., Gaudenz, K., Jay, P. *et al.* (1996). Mutations in the human *Sonic Hedgehog* gene cause holoprosencephaly. *Nature Genetics*, **14**, 357–359.

Stuart, E.T. and Gruss, P. (1995). PAX genes: what's new in developmental biology and cancer? (Review). *Human Molecular Genetics*, **4**, 1717–1720.

Watnick, T.J., Gandolph, M.A., Weber, H., Neumann, H. *et al.* (1998). Gene conversion is a likely cause of mutation in *PKD1*. *Human Molecular Genetics*, 7(8), 1239–1243.

Wengler, G., Gorlin, J.B., Williamson, J.M., Rosen, F.S. *et al.* (1995). Nonrandom inactivation of the X chromosome in early lineage hematopoietic cells in carriers of Wiskott–Aldrich syndrome. *Blood*, **85**(9), 2471–2477.

Wilkie, A. (1997). Craniosynostosis: genes and mechanisms (Review). *Human Molecular Genetics*, **6**(10), 1647–1656.

Self-assessment questions

1. Give examples of two different types of mutations and briefly explain their effects on gene function.
2. What is a triplet repeat? How might a clinical disorder arise from a triplet repeat? Give an example.
3. In the fragile X syndrome, the expansion of triplet repeats becomes methylated. The affected boys do not produce the FMR-1 protein. Why?
4. A female with a gene mutation on the X chromosome for the biochemical disorder Hunter syndrome develops that disease despite the presence of a second normal X. What process may be occurring?
5. A baby is born with a particular kind of hydrocephalus (fluid collecting in the brain). When the chromosomes are analysed, it is found that both chromosome 14s come from the mother. Name this phenomenon and explain why the presence of the father's 14 is required for a normal phenotype.
6. An individual's karyotype reveals a balanced chromosome translocation, and yet the patient shows clinical symptoms. It is noted that one of the breakpoints lies very close to a centromere. Why might this be significant?
7. In a famous experiment on the fruit fly *Drosophila* it was observed that instead of antennae growing on the head of the fly there was a leg instead (antennapaedia). Given that the orthologous human genes have a similar function, which group of developmental genes is likely to be involved?

Key Concepts and Facts

Characterization of DNA Mutations

- DNA mutations can be classified into categories such as base substitutions, missense, nonsense or frameshift mutations. Such mutations lead to loss or gain of function. Triplet repeats are dynamic mutations which may produce a clinical effect by expansion.

Alteration of Normal Gene Control Mechanisms

- Gene mutations can also alter the control of gene expression through various mechanisms such as positional change, haplo-insufficiency and inappropriate methylation, leading to loss or gain of function. This may occur on a larger scale through chromosomal changes resulting in aneuploidy, polyploidy or structural rearrangements such as translocations or duplications.

Alteration of Immunogenetic Expression

- Mutations in genes responsible for the formation of B and/or T cells lead to the inability to produce mature or functioning B or T cells, so that immunity against infection is compromised.

DNA Repair Syndromes

- DNA syndromes display increased rates of cancers and immune disorders. Mutations of excision repair genes, mismatch repair genes or genes involved in DNA replication lead to uncorrected base changes and failure of DNA repair.

Abnormalities of Developmental Genes

- Mutations of the *FGFR* growth receptor genes may result in craniosynostosis or dwarfism. Mutations in *SHH* and *PTCH* result in abnormal cell movement leading to facial, gut or lung abnormalities or cell overgrowth. Homeobox mutations lead to wrongly positioned or abnormal numbers of digits or limbs.

Abnormalities of Sex Determination

- The 46,XY genotype may be expressed as a female phenotype due to mutations or deletions of the *SRY* gene, or a defect in the androgen receptor gene. A 46,XX genotype may appear phenotypically male due to the presence (usually on one of the X chromosomes) of the *SRY* gene, or due to hormonal abnormalities such as those seen in CAH.

Part Two:
Diagnosis of Disease

Chapter 4
Patterns of inheritance

Learning objectives

After studying this chapter you should confidently be able to:

List Mendel's two laws of inheritance.

Outline the principles of linkage and recombination.

Draw simple pedigrees using the correct nomenclature and symbols.

Describe the three major inheritance patterns and give examples of appropriate disorders.

List other problems causing deviations from Mendelian inheritance patterns.

Define the Hardy–Weinberg equation in terms of the equilibrium of a population, and list factors disturbing that equilibrium.

Explain how LOD scores can be used in linkage studies.

Describe how Bayes' theorem is used in assessing carrier risks.

Classical genetics

Mendel's laws

The basis for the understanding of inheritance came from the classic work on plants by Gregor Mendel. By means of meticulous breeding experiments with peas, Gregor Mendel explained the basic principles of genetic inheritance, which we now accept to be applicable to all higher organisms with a sexual method of reproduction. Those concepts of heredity, forming the whole basis of the science of genetics, are now commonly referred to by the terms 'Mendelism' and 'Mendelian inheritance'.

As Mendel did not know about chromosomes or genes, he used the term **'factor'**.

When considering a factor such as the height of pea plants, for example, he identified the pure **characters** (or traits) 'tall' and 'short', which in modern terminology would now be described as **alleles** (alternatives) of a gene coding for height.

Gregor Mendel was an Austrian monk whose classic experiments used seven sets of well-defined paired characteristics exhibited by pea plants. He first published the data from his experiments with peas in 1866; the full significance of his two laws to genetics was not appreciated until 1900.

The following terms are defined with respect to modern human genetics.

- **Allele:** One of several alternative forms of a gene at a specific gene locus. An individual will therefore have two alleles (one maternal, one paternal) at each autosomal locus.
- **Heterozygous:** Having two different alleles at a specific gene locus.
- **Homozygous:** Having two identical alleles at a specific gene locus.
- **Hemizygous:** Having only one copy of a gene (and therefore one allele); for example on the single X chromosome in a male, who is therefore hemizygous for X-linked genes.
- **Dominant:** A genetic trait which is observed (or expressed) in the heterozygous state and refers to the effect of an allele or 'character' at a specific gene locus.
- **Recessive:** A genetic effect of an allele that is manifest in the homozygous state at a specific gene locus.
- **Genotype:** The genetic contribution of an individual (or cell). A genotype may also refer to the constitution of alleles at a given locus in an individual.
- **Phenotype:** The observable physical effect on an individual (or cell), determined by the genetic constitution.

Following the hybridization of two pure traits, Mendel termed characters which were transmitted 'entire or almost unchanged' **dominant**, and those which became 'latent' as **recessive.**

Mendel's conclusions are most simply expressed as two **principles:** firstly that of **segregation** and secondly that of **independent assortment.**

Mendel's first law: The principle of segregation

- **Factors come in pairs and have different characters or traits.** Modern interpretation: One gene is inherited from each parent and hence alleles are inherited in pairs.
- **Only one character is passed on from each parent.** Modern interpretation: Although there may be many alleles at a given gene locus, following meiosis only one allele enters the gamete.
- **Equal numbers of gametes inherit each allele.**

Mendel's second law: The principle of independent assortment

Following experiments using unrelated pea plant characteristics such as pea colour and pea shape, Mendel proposed that:

- **Different characters assort independently.** As long as the genes for colour and shape are on non-homologous chromosomes, the alleles (such as yellow/green or smooth/wrinkled) segregate independently at meiosis.

Linkage

Mendel's second law only works as long as genes are on separate chromosomes, or far enough apart on the same chromosome to assort independently. The closer a pair of genes are (on the same chromosome arm for example), the less likely there is to be independent assortment – that pair of genes tend to be inherited together, and are said to be **linked.** Whole blocks of genes may be inherited together – the composite genotypes are then known as a **haplotype.**

Recombination

During meiosis there is at least one crossover (chiasma) between every pair of homologous chromosomes. Sometimes on longer chromosomes there may be more than one chiasma, as there is more room (and hence more chance) for this to occur (Figure 4.1).

Although genes on different chromosomes segregate independently, if genes are well separated on the same chromosome there

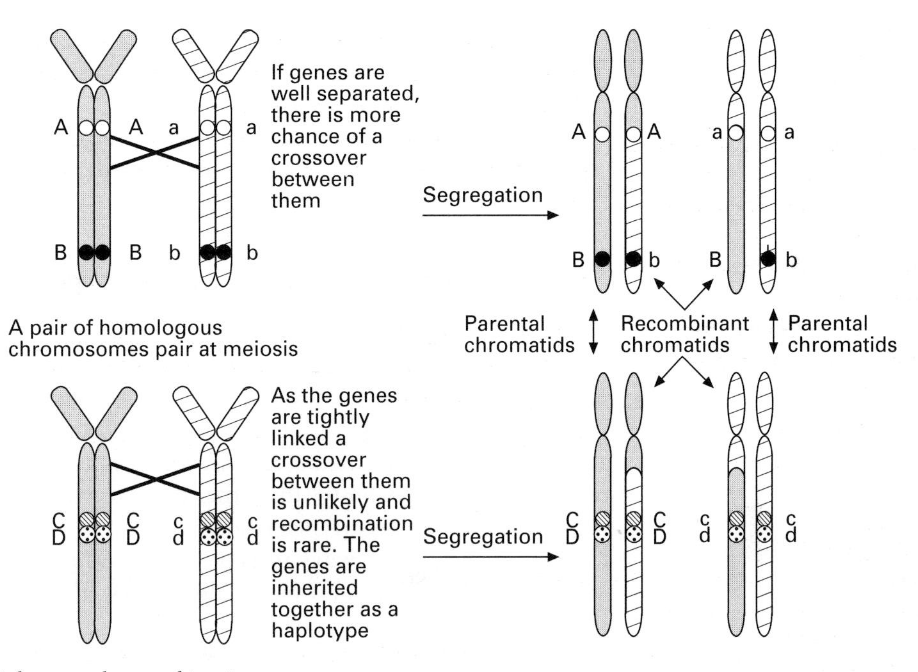

Figure 4.1 *Linkage and recombination*

is still a good probability that a crossover will result in independent segregation. This produces 2/4 parental alleles and 2/4 **recombinants**, i.e. a recombination rate of 50%.

This is sometimes called a **recombination fraction** (also known by the Greek letter theta, θ), such that a recombination rate of 50% is expressed as 0.5.

The closer genes are on the same chromosome, the more closely **linked** they will be, and there will be less chance of a crossover occurring between the gene loci, with fewer recombinants and a lower recombination fraction.

Genes are sometimes said to be syntenous. **Synteny** means that the genes are on the same chromosome, though not necessarily displaying genetic linkage with each other (i.e. they assort independently). They are known to be on the same chromosome because they are both linked to intermediate genes.

The application of Mendel's laws to human genetics

Although Mendel formulated his laws using data derived from pea plants, exactly the same principles apply to inheritance patterns in humans. Detailed knowledge and understanding of the properties of human genes was harder to obtain simply because it was not possible to undertake the same sorts of breeding experiments that could be carried out on peas, fruit flies and fungi.

To begin with, knowledge and understanding of human genes was drawn mainly from constructing family pedigrees, sometimes to the extent of searching through historical archives and parish records, and subjecting the data to complex statistical analysis.

It is only within the past 20 years, since the development of modern techniques of molecular biology, that substantial progress has been made in terms of understanding the intricacies of human genetics, but deduction of basic patterns of inheritance often still requires pedigree analysis.

Around 5000 human diseases are now traceable to mutations in single genes. These are the **single gene disorders** and the pattern of their inheritance is often well characterized.

Pedigrees

A family with a genetic disorder will usually be referred to a **clinical geneticist,** a medical consultant with specialist training in the identification of genetic diseases. It is important to elucidate the **mode of inheritance** of the disease, in order to give the family an idea of the probability that the disease will recur in future generations – the **recurrence risk.**

One of the easiest ways to start this investigation process is to take a family history and pictorially express this in the form of a family tree, known as a **pedigree.**

The symbols used in such pedigrees can be seen in Figure 4.2. Each generation (grandparents, parents and children for example) is

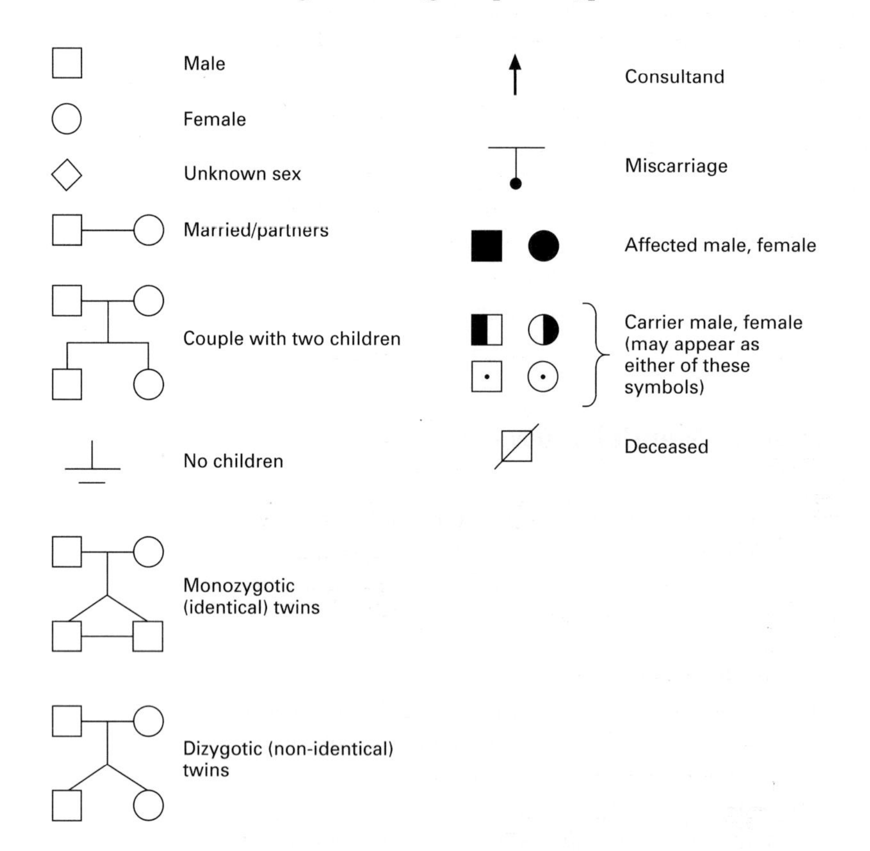

Figure 4.2 *Pedigree symbols*

given a Roman numeral (I, II, etc.). Each individual within that generation is sequentially numbered from left to right using Arabic numerals. Miscarriages and pregnancies are also included.

The affected individuals are indicated on the pedigree; by studying how frequently the disease was passed on and in which sex, the geneticist can often deduce the **pattern of inheritance.**

Mendelian inheritance patterns

There are three major and two less common types of inheritance pattern (Figure 4.3). These are classified according to whether the gene responsible for a particular characteristic or disorder resides on a sex chromosome or an autosome, and also whether that gene is expressed in its homozygous or heterozygous state.

Autosomal dominant (AD)

Autosomal dominant inheritance follows the general rules given below:

- A dominant gene is found on one of a pair of homologous **autosomes.** This means that either the father or the mother can pass on a dominant disorder.
- Only **one** parental gene is required to display the associated phenotype – it is **dominant** to the allele on the homologous chromosome. Although two alleles are present at that gene locus (the individual is heterozygous), the dominant allele is able to be **expressed in this heterozygous state.**

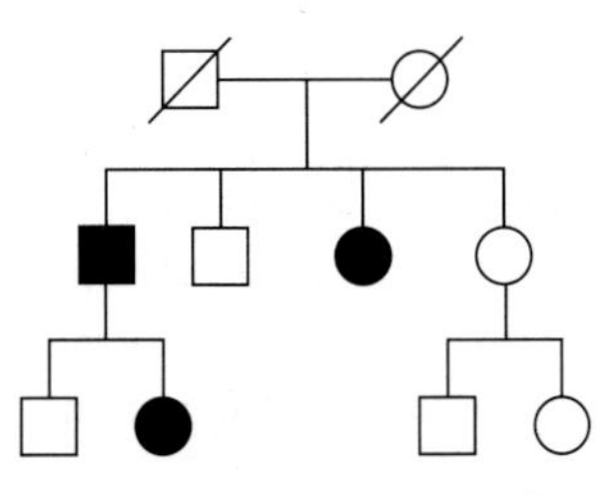

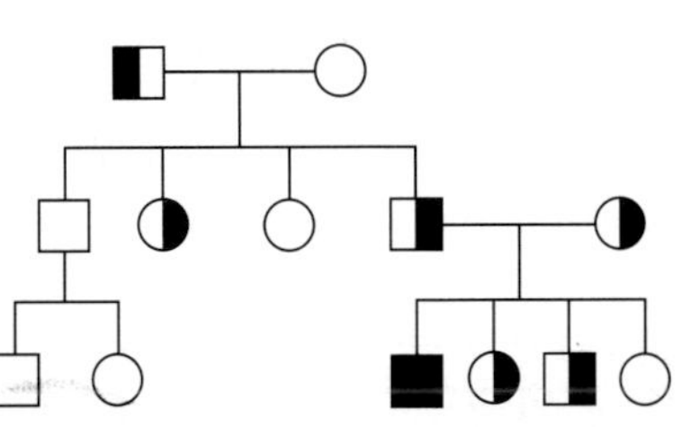

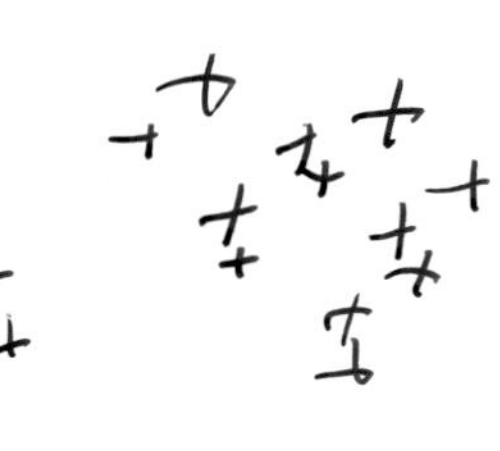

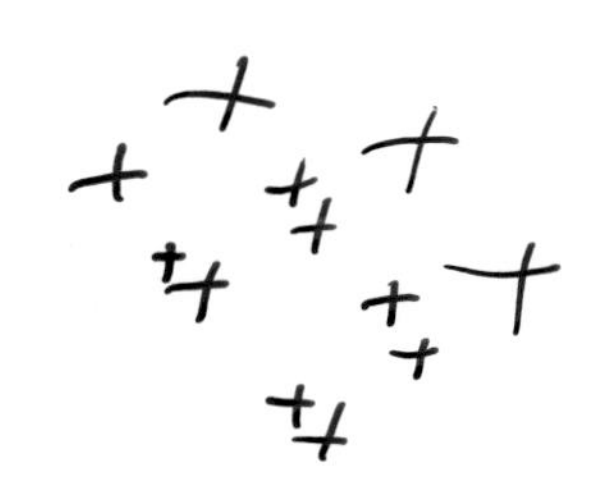

X-linked recessive (XLR)

Figure 4.3 *Inheritance patterns*

Table 4.1 *Examples of autosomal dominant genetic diseases*

Disease	*Chromosome locus*
Huntington disease	4p16.3
Myotonic dystrophy	19q13.3
Adult polycystic kidney disease type 1	16p13.1
Waardenburg syndrome type 1	2q35
Neurofibromatosis type 1	17q11
Familial adenomatous polyposis coli	5q21

- The children of a parent with an autosomal dominant disorder have a **50% chance** of inheriting that disorder, **irrespective of sex.** If an individual is phenotypically (and hence genotypically) unaffected, his or her children will also be unaffected.
- An autosomal dominant **pedigree** has a high chance of having affected family members in each generation, and is sometimes said to have a '**vertical**' appearance (Figure 4.3).

Some examples of autosomal dominant genetic diseases are given in Table 4.1.

Other features of autosomal dominant disorders

Co-dominance

This occurs when the characteristics of each of two alleles are expressed in the heterozygous state.

The most well-known examples are the ABO blood groups, such that a person with blood group AB shows antigens for both group A and group B.

Reduced penetrance

In most autosomal dominant diseases, a certain **proportion of individuals** will not show the expected abnormal phenotype, sometimes to the extent of appearing completely phenotypically normal (non-penetrant). Depending on other factors such as environment, the proportion for specific diseases may be predictable.

For example, in retinoblastoma (a childhood tumour of the eye) the penetrance is 90%, such that in any 10 individuals carrying the *RB1* gene, one will not develop the tumour. The mechanism of this particular disorder is now well understood and is described in Chapter 8.

Biochemical disorders such as porphyria and hereditary pancreatitis are examples where certain carriers may only show biochemical but no overt external symptoms.

Variable expression

This refers not to the individual in isolation, but the observation that there can be a great deal of phenotypic variation between different members of the same family, who all carry the same mutation in a dominant gene. This variation is the **degree to which the gene is expressed.**

A uniform degree of expression is the exception in autosomal dominant disorders. Achondroplasia (dwarfism) is one of the few disorders in which the heterozygous state usually results in a typical recognizable phenotype.

In contrast, affected members of a family with Waardenburg syndrome may show different combinations of the characteristic phenotypic features such as a white forelock, deafness and unusual eye pigmentation.

Neurofibromatosis type 1 (NF 1) may result in either a nearly normal phenotype comprising a few café au lait spots, which look a little like freckles, through to many large benign tumours on the external surface of the skin.

Autosomal dominant disorders with extreme variation in expression result in the apparent phenomenon of 'skipped generations'. The abnormal gene is passed through every generation, but one key individual may have been so mildly affected that he or she appeared to be normal.

Problems arise in predicting the phenotype of an unborn child who has inherited the abnormal gene. A parent who is very mildly affected with NF 1, for example, has to be made aware of the full range of abnormalities that may be expressed in their child, and that they may be considerably worse than those of the parent.

Late onset

The most notable example of a late onset disorder is Huntington disease (HD). This neurological disorder affects the brain, leading to severe motor disturbances, both physical and mental. It is often not expressed until the third or fourth decade of life, when an affected individual may have already had a family. Each of those children will then have a 50% risk of having inherited HD.

Anticipation

Although anticipation is found in other types of inheritance patterns (see the section on X-linked inheritance for fragile X), some of the original examples were noted in autosomal dominant diseases. As the faulty gene is passed on, the phenotype appears to become more severe (or exhibit more clinical symptoms) in later generations.

Myotonic dystrophy, for example, may show minimal symptoms in a grandfather, who might have a slight drooping of the eyelid

Myotonia is an inability to relax muscles. Patients have a typical handshake; once they have gripped the clinician's hand they find it very difficult to let go.

muscles, baldness, cataracts, but a normal lifespan. His daughter may have a more pronounced weakening of the facial muscles including the typical downturned mouth and hand myotonia. This disease is usually more severe if it is passed through the mother, and her newborn child may then have congenital myotonic dystrophy, which results in a severe whole body muscle weakness, mental retardation and poor survival.

The underlying molecular genetic basis for this is now known, and is due to a number of triplet repeats (CTG) on chromosome 19, which expand from generation to generation until a critical number is reached, resulting in a severe phenotype (see Chapters 3 and 7).

Autosomal recessive (AR)

Autosomal recessive inheritance follows the general rules given below.

- A recessive gene is found on **both** of a pair of homologous **autosomes.** In other words, the father and mother have each passed on a recessive gene to their child.
- Both parental genes are required to display the associated phenotype, which is only revealed when the mutant alleles are in the **homozygous state.**
- The children of parents each heterozygous for an autosomal recessive disorder have a **25%** chance of being **affected,** a **50%** chance of being a phenotypically unaffected **carrier,** and a **25%** chance of being **normal, irrespective of sex.** A healthy sibling of a known affected child has a 2/3 chance of being a carrier.
- An autosomal recessive **pedigree** has less chance of having affected members in every generation, as any carriers would have to meet another carrier with the same defective gene from the general population. These pedigrees have a **'horizontal'** look, or even just one apparently 'sporadic' case (Figure 4.3). This is especially true in small families.
- These pedigrees also give rise to the notion of 'missing a generation', when a carrier may not have any affected children, but their phenotypically normal children may be carriers who will meet another carrier in the next generation.

The reason why a healthy sibling of a child affected with an autosomal recessive disorder has a 2/3 chance of being a carrier is as follows:

- The children of heterozygous parents have genotypes in the proportion **1 normal** : **2 carrier** : 1 affected.
- As the phenotypically normal sibling is not affected, the affected genotype is not relevant to the calculation.
- The proportion of normal phenotypes is therefore 1 normal : 2 carrier.
- Therefore the carrier risk to the healthy sibling is 2/3.

Some examples of autosomal recessive disorders are given in Table 4.2.

Other features of autosomal recessive disorders

Consanguinity

In order for an affected child to be born, the parents must be two asymptomatic carriers of the same disorder. The chance of them

Table 4.2 *Examples of autosomal recessive disorders*

Disease	*Chromosome locus*
Cystic fibrosis	7q31
Phenylketonuria	12q22-24.1
Spinal muscular atrophy type 1	5q13
Tay–Sachs disease	15 and 5 (code for two subunits of the enzyme hexosaminidase A)
Congenital adrenal hyperplasia	6p

meeting is dependent on the frequency with which that gene mutation is found in the general population.

Usually for most autosomal recessive conditions the gene frequency is low, but if a couple are genetically related, the presence of a recessive mutation in a common ancestor make it more likely that they may **both** have inherited the mutation.

The exact risk that a related couple both carry the same inherited mutation depends on their genetic relationship and hence the number of genes they would share. First cousins share 1/8 of their genes; if their common grandparent passed on a mutation, the chance that their child will inherit both mutations is 1/32 – despite the fact that the frequency of the disease may be very low (say 1/40 000 for a typical biochemical disorder) in the general population.

For a rare disease, a gene frequency of 1/200 in the population would result in a carrier frequency of about 1/100 and hence a disease incidence of 1/40 000 (see the Hardy–Weinberg equilibrium later in this chapter).

However, some disorders have a much higher gene frequency and hence a higher population carrier risk. Examples are cystic fibrosis, with a British carrier risk of about 1/24, and haemochromatosis (a treatable disorder of iron metabolism) which has a carrier risk of 1/10.

Each cousin has a carrier risk of 1/4 from the common grandparent. The chance of them both passing this on to a child is 1/2 each; that is $(1/4 \times 1/2) \times (1/4 \times 1/2) = 1/64$. Because they share **two** common grandparents the risk is twice as great, i.e. $2 \times 1/64 = 1/32$.

X-linked recessive (XLR)

X-linked recessive inheritance follows the general rules given below.

- An X-linked gene is found on the **X chromosome.**
- **Males** who receive an abnormal X from their mother will be **affected.**
- **Females** who carry one normal X and one abnormal X will usually be **unaffected** carriers.
- Affected male children are usually born to unaffected parents; the father (who contributes his Y chromosome) will usually be normal and the mother will be an asymptomatic carrier. Such a couple will have a **25%** risk of an **affected son,** a **25%** risk of a **normal son, 25%** risk of a **carrier daughter** and a **25%** risk of a **normal daughter.**
- An X-linked **pedigree** is characteristic in that the **affected individuals will all be male** (Figure 4.3). Female carrier status may be inferred in a woman with both an affected brother and son.

Some examples of X-linked recessive disorders are given in Table 4.3.

Table 4.3 *Examples of X-linked recessive disorders*

Disease	*Chromosome locus*
Duchenne muscular dystrophy	Xp21
Becker muscular dystrophy	Xp21
Haemophilia A	Xq28
X-linked ichthyosis (steroid sulphatase)	Xp22.3
Adrenoleucodystrophy	Xq28
Hunter syndrome (MPSII)	Xq28

Other features of X-linked recessive disorders

Reproductive ability of the affected male

Normally an X-linked disorder is passed on by a phenotypically healthy female carrier. Some X-linked diseases result in a lethal outcome for an affected male before the age of reproduction, so there is no chance of his abnormal X being passed on. This is the case with Duchenne muscular dystrophy and the severe form of Hunter syndrome.

However, some disorders are now treatable, or are mild enough to allow the affected males to marry and reproduce. Haemophilia A can now be treated with infusions of Factor VIII. Becker muscular dystrophy (which is an allelic form of Duchenne muscular dystrophy) can have a very mild phenotype.

If an affected male can reproduce (for example in haemophilia), **all his daughters will be carriers**, as they receive his single abnormal X (together with a normal maternal X). **All his sons will be normal**, as they will receive his normal Y together with their mother's normal X.

X-linked dominant inheritance (XLD)

X-linked dominant inheritance is rare, but follows the general rules given below.

- An X-linked dominant gene is carried on the **X chromosome.**
- Both **males and females** will be **affected.**
- The **females** may have a more **variable expression** as they also have a normal X (the phenotype may depend on the randomness of the X-inactivation).
- If a **female is affected, each child** will be at **50% risk of being affected**, irrespective of sex. If a male is affected, all his daughters (but none of his sons) will be affected.

Some examples of X-linked dominant disorders are given in Table 4.4.

Table 4.4 *Examples of X-linked dominant disorders*

Disease	*Chromosome locus*
X-linked dominant retinitis pigmentosa	Xp11
Incontinentia pigmenti (IP)	?Xq28 (familial). Gene locus uncertain
Rett syndrome	Xq28

Fragile X – a special case?

Fragile X was originally thought to be an X-linked recessive disorder, until it was found that around 1/3 of obligate carrier females had clinical symptoms. This is still too low a percentage to be a fully penetrant X-linked dominant disorder (all female carriers would then be affected).

As explained in Chapter 3, the molecular basis of fragile X is now known to involve a DNA triplet repeat CGG, which increases down the generations until a critical number of repeats are reached (usually around 200). At this point the gene becomes methylated, such that it cannot produce the protein FMR-1, with the consequent phenotypic results (Figure 4.4). The disease therefore appears to display **anticipation**.

Y-linked inheritance

There are no well-characterised clinical disorders found on the Y chromosome. There are very few Y-linked characteristics, but they follow the general rules of inheritance given below.

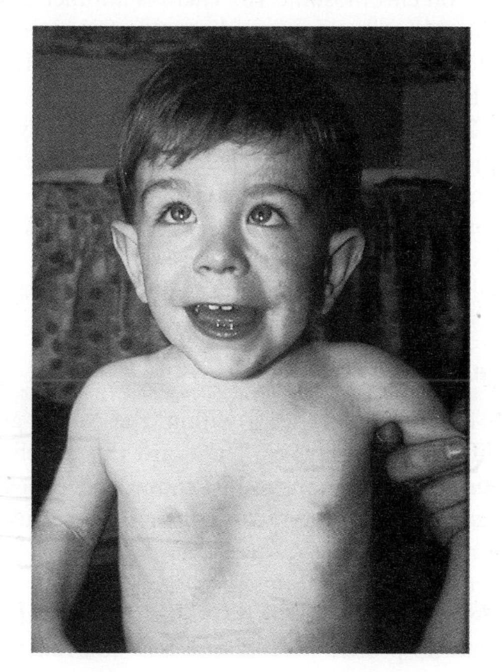

Figure 4.4 *Fragile X syndrome*

Table 4.5 *Examples of Y-linked disorders/characteristics*

Disorder/characteristic	*Chromosome locus*
Sex determining region on Y (*SRY*)	Yp
Azoospermia	Yq11.23

Locus heterogeneity can be detected when a couple who each appear to have the same disorder marry and have normal children. Deafness is one well-known example where several different genes may interact in the complex processes leading to the normal function and anatomy of the ear.

A deaf couple each may have a mutation on a different chromosome and a resulting normal child may in fact be heterozygous for each autosomal recessive form of deafness at two unrelated gene loci.

When genes react in this way they are said to be **complementary**; the interaction of two non-allelic gene products such that a normal phenotype is produced is called **complementation**.

In certain circumstances the interaction of gene products may be demonstrated by mixing cell lines from two individuals and revealing the presence of a functional protein.

The corollary of complementation is the **compound heterozygote**. In this case, each parent carries a different intragenic mutation which is passed on to the child, but there is no complementation and the child is affected.

An example would be cystic fibrosis, where any mutation in the *CFTR* gene usually leads to disruption of the protein function. The child inherits two different CF mutations, but the two proteins are dysfunctional in different ways and do not complement. Thus CF behaves in a conventional recessive manner.

- Y-linked characteristics are found on the **Y chromosome.**
- All male children would inherit that characteristic.
- If a male did have a Y-linked clinical disorder, **all his sons would inherit it.**

Two examples of Y-linked disorders/characteristics are given in Table 4.5.

Other problems with inheritance

Heterogeneity

Locus heterogeneity

There are two different kinds of heterogeneity. If a mutation in a gene produces a specific phenotype, and a mutation of a completely **different gene** (perhaps on a different chromosome) produces a **similar phenotype**, the disease is said to show **locus heterogeneity.**

Example 1: There are several kinds of polycystic kidney disease. Adult polycystic kidney disease type 1 (APKD1) is dominant; the gene is found on chromosome 16. There is another dominant form (APKD2) with virtually the same phenotype found on chromosome 4. A third form is infantile autosomal recessive polycystic kidney disease (ARPKD), found on chromosome 6.

Example 2: There are many kinds of muscular dystrophies and muscular atrophies. There are twelve types of spinocerebellar ataxia for example, all found on different chromosomes.

It is important that the clinician correctly identifies the locus of that family's disease correctly, otherwise DNA tests, which depend heavily on the correct gene location, will be inaccurate.

Allelic heterogeneity

Sometimes different mutations within the **same gene** produce clinically **different phenotypes.** These are then **allelic forms**, one example being the severe phenotype found in Duchenne muscular dystrophy, compared to the much milder Becker muscular dystrophy. These are found at the same gene locus (Xp21) but tend to have different allelic mutations. Although both diseases may be caused by deletions, those in DMD destroy the reading frame of the gene such that no protein (or a completely non-functional protein) is produced. BMD has in-frame deletions which result in a smaller protein with the functional ends intact.

Parental origin

In Chapter 2, we saw that parental **imprinting** played a natural part in some forms of gene expression. Inheritance of some normal gene functions may therefore depend on the **parent of origin**, as demonstrated by diseases such as Prader–Willi syndrome, Angelman syndrome and Beckwith–Wiedemann syndrome.

New mutations

Not all mutations are inherited. Some are **new mutations**; they arise in the parental germ cells and appear for the first time in the affected individual.

In those diseases that are **lethal before reproductive age**, the generation of new mutations serves to replace the old mutations leaving the population. In lethal dominant or X-linked diseases, therefore, the **new mutation rate tends to be fairly high.**

In Duchenne muscular dystrophy, for example, approximately 1/3 of all affected boys result from a new mutation on the mother's X chromosome.

In diseases where affected individuals live to reproduce (for example the **late onset** Huntington disease), the **natural mutation rate is low**, as the inherited mutation is maintained at a higher level in the population.

Germline mosaicism

Occasionally a small group of mutant cells may arise in the male or female germ cells as a clone. Although normal germ cells are also present, one of the mutant germ cells may be inherited, resulting in an affected child.

For example, if a woman has two sons with Duchenne muscular dystrophy and she has no other family history, there are two possibilities:

- She may be a carrier as she inherited an abnormal X from her mother.
- She could be a germline mosaic, the mutation having arisen in her germ cells.

The carrier risks to various members of the family are very different with these two scenarios. The sisters of a germinal mosaic would have a low carrier risk, but if their mother is an obligate carrier, they would each be at 50% risk of being carriers. Extended studies of the family history and molecular genetics studies in the laboratory can be undertaken to determine carrier risks.

As there is cooperation between the nuclear and mitochondrial genomes (see Chapter 1), some mitochondrial diseases are not maternally inherited, but show a Mendelian pattern. Examples are Leigh syndrome, which is autosomal recessive in 7–20% of cases, and autosomal dominant progressive external ophthalmoplegia (adPEO).

As mitochondria are maternally inherited, it was theorized in the 1980s that a female ancestor could be traced back over 200 000 years; she would have been the 'Eve' of the present day population.

Although the conclusion of the example above may be exaggerated, mitochondrial DNA has been used, for example, in proving the identity of the remains of the Russian royal family (the Romanovs) who were shot after the Russian Revolution in July 1918. Their mtDNA was compared to that obtained from a blood sample given by Prince Philip, the Duke of Edinburgh, who shared a common female ancestor.

Height was thought to be typically polygenic until the discovery of the *SHOX* gene on the X chromosome, which probably accounts for around 70% of our final height. There are, however, other genes for height, one of which is thought to be located on Yq.

Mitochondrial inheritance

As the cytoplasm of our cells is derived from the cytoplasm of the maternal egg cell, **mtDNA is inherited from the mother by cytoplasmic inheritance**. Sperm have very little cytoplasm, and have very few mitochondria. For this reason a fertilized cell has both a maternal and paternal set of chromosomes, but only maternal mitochondria. It therefore follows that if mutations occur in the mitochondrial genome, they will be passed to children of both sexes by the mother.

Sometimes, however, a particular mutation is not present in every mitochondrion. There will then be a mosaic population comprising normal and abnormal mitochondria, known as **heteroplasmy.** Where heteroplasmy exists, the proportion of normal to mutated mitochondria may determine the severity of the disease, and this is known to be variable in mitochondrial disorders.

Examples of mitochondrial disorders are given in Chapter 3.

Gradation of inheritance

The most straightforward patterns of inheritance are found in the **Mendelian disorders**, which are usually based on the premise of 'one faulty gene, one disorder'.

Chromosomal abnormalities may be incorporated into the parental germline, and will then follow Mendelian patterns of segregation.

Some characteristics, however, such as skin colour or intelligence, depend on the additional small effects of each of a number of genes. These are called **polygenes.**

Prediction for such characteristics is therefore much more uncertain. When the effects of environment are also taken into consideration, it becomes very difficult to isolate the genetic from the environmental components of a disease.

There are certain disorders which appear to be more likely to express themselves if the individual has certain genetic markers known to be associated with the disease.

These are **susceptibility** genes or markers (Table 4.6), which may sometimes need an extra environmental factor in order to develop the disease phenotype. Not everyone with the marker DQA1 will develop multiple sclerosis, for example; the other factors necessary

Table 4.6 *Susceptibility markers*

Disorder	*Gene/marker*	*Allele*
Ankylosing spondylitis	HLA (tissue) type	B27
Multiple sclerosis	HLA (tissue) type	DQA1
Alzheimer disease	Apolipoprotein	E4
Rheumatoid arthritis	HLA (tissue) type	DR4

to instigate the disease may range from viruses to a faulty immune system or even diet.

There is no true distinction between 'polygenes plus an environmental effect' and the term **multifactorial disorders**, which was the name formerly applied to diseases with no clear inheritance pattern but with an element of genetics.

Heart disease covers a multitude of different heterogeneic anatomical and phenotypic problems, yet we now know of a particular susceptibility if an individual carries the gene for familial hypercholesterolaemia. The males are prone to heart attacks in their thirties and forties, yet if they can follow a suitable diet from childhood and have cholesterol-lowering drugs, this risk is considerably lowered.

Identical twins, who have identical genotypes (they are essentially clones), are a naturally occurring phenomenon that allows us to study the different effects of nature (genetics) and nurture (environment).

Although genetically identical, their environmental experiences may be quite different. There are cases of twins being separated and adopted at birth into different family environments. By noting whether one twin develops a particular disorder and then observing if the other one also develops the same disease (**concordance**), we are able to estimate the genetic contribution towards that disease compared with the environmental effect.

Population genetics

Each gene may have two or more alleles, which may express themselves as harmless polymorphisms (differences) such as eye colour or unimportant DNA sequences in introns; alternatively they may represent a mutant gene on one chromosome and a normal gene on the homologue.

In a large heterogeneous population many alleles will be represented. Normal and abnormal alleles will be present in the heterozygous and homozygous form.

There appears to be a natural balance of alleles in large populations; they are in **equilibrium**.

The Hardy–Weinberg equilibrium

The Hardy–Weinberg equilibrium is a mathematical way of representing the balance of alleles in a population. Each **allele** (note that these are alleles of genes, not the genes themselves) is present in a population at a particular **frequency** (Figure 4.5).

If we take a gene that has two alleles **A** or **a**, then there are three combinations possible at the gene locus on a pair of autosomes: **AA**, **Aa** or **aa**.

If we say that **A** has an allele frequency of **p** (i.e. A = p) and that **a** has an allele frequency of **q** (i.e. a = q) and that in a population all the alleles must add up to **100%** (i.e. 100% = **1**), then:

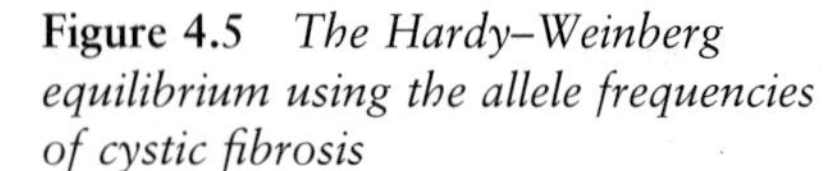

Figure 4.5 *The Hardy–Weinberg equilibrium using the allele frequencies of cystic fibrosis*

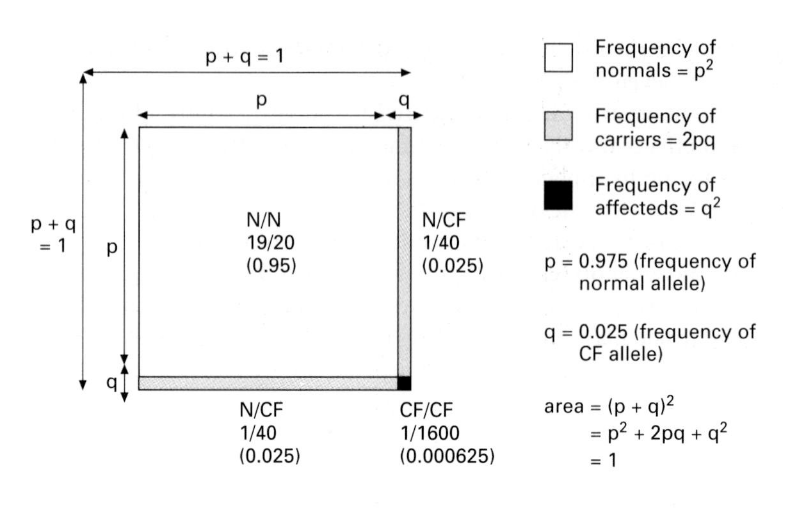

If $\mathbf{p+q=1}$, then $(\mathbf{p+q})^2\mathbf{=1}$, so expanding the equation gives:

$$p^2+2pq+q^2=1$$

Let this be illustrated by the recessive disease cystic fibrosis.

If **A** = the normal allele, then the **normal allele frequency** = **p**, and if **a** = the CF (abnormal) allele, then the **abnormal allele frequency** = **q**.

Then: $\mathbf{p^2}$ = the proportion of **normal** people in the population; **2pq** = the proportion of **carriers** in the population; and $\mathbf{q^2}$ = the proportion of CF **affected** people in the population.

We can use the Hardy–Weinberg equation to calculate various useful pieces of information such as carrier rates and incidence of the disease.

If the **incidence** of newborns with CF in the Caucasian population is **1/1600**, then $q^2 = 1/1600$ and so $\mathbf{q = 1/40}$ (0.025), which is the **CF gene frequency** (or abnormal allele frequency).

As $\mathbf{p=1-q}$, then $\mathbf{p=0.975}$ (1−0.025), which is the **normal gene frequency** (≈1) and:

$$\mathbf{2pq} = 2 \times 0.975 \times 0.025$$
$$= 0.04875 = \mathbf{1/20}$$

which is the **carrier** frequency.

$$p+q=1$$

This also means that $\mathbf{p=1-q}$ and that $\mathbf{q=1-p}$.

However, the Hardy–Weinberg equation describes the **distribution of alleles in a population**, i.e. how many AAs (i.e. $\mathbf{p \times p = p^2}$), how many **Aas** and **aAs** (i.e. $\mathbf{2 \times p \times q = 2pq}$) and how many **aas** (i.e. $\mathbf{q \times q = q^2}$).

When numbers are squared in this way, the allele distribution is represented by a **binomial equation**, which also shows the balance (**equilibrium**) of the three combinations of alleles:

$$p^2+2pq+q^2$$

Factors disturbing the Hardy–Weinberg equilibrium

The Hardy–Weinberg equation only applies to large, randomly mixed populations. It depends upon a small number of abnormal genes disappearing from the population and being replaced at a constant rate. Physical or geographical factors may therefore influence the level of an abnormal gene or unusual polymorphism in a population, as outlined below.

Random genetic drift

A small number of people may become separated from the main population either voluntarily (by emigration) or involuntarily (due to features of the terrain such as mountains, rivers, etc.).

There is a chance that in a small population (who may be more closely related anyway) an imbalance of genes may exist, such that there are more abnormal genes for a particular disease than expected according to the Hardy–Weinberg equation.

If an imbalance in a clinically insignificant polymorphism occurs, it may increase in frequency (by chance due to genetic drift) to become the only allele carried in that population (American Indians do not have blood group B for example).

Sometimes the origin of a particular gene abnormality can be can be traced back through the ancestors of the affected family members to a single individual.

There is only one mutation for the second breast cancer gene (*BRCA2*) in Iceland. Researchers at the Icelandic Cancer Society discovered that a sixteenth century cleric called Einar was responsible for nearly every case of this type of breast cancer in Iceland today. It was possible to trace him, as historically there was a relatively small gene pool on this remote island.

The individual bearing the first mutant gene found in a population leads to the phenomenon of **the founder effect.**

Another way of considering carrier risk is to say that each chromosome 7 (where the CF gene locus is found) has a 1/40 risk of carrying CF. As there are two chromosome 7s, the risk is $1/40 + 1/40 = 1/20$.

If we just know the carrier risk, the incidence is calculated as follows.

The chance of a female (carrier risk 1/20) meeting a male (carrier risk 1/20) and having an affected child (affected risk 1/4) is **$1/20 \times 1/20 \times /4 = 1/1600$.**

Change in mutation level

The stability of allele frequencies will alter in a population if there is a change in mutation level, resulting in an increase or decrease of q^2.

An increase or decrease in population size may influence the frequency of an allele. If a population becomes smaller, for example due to illness, an abnormal allele may remain in the survivors at a higher frequency than previously.

This allele will be perpetuated as that population grows again. With any inbred population there is an increased risk of autosomal recessive disorders.

An abnormal allele can sometimes be maintained at an abnormally high level due to the phenomenon of **heterozygote advantage.** Here the **carrier state for a particular autosomal recessive disorder** confers a physical advantage over the homozygous normal state and is selected for.

As described in Chapter 3, sickle cell anaemia is common in countries where malaria is found. Although HbS red cells carry oxygen normally, in the deoxygenated state the red blood cells are shaped like sickles. Sickled cells are poorly deformable and cannot traverse capillaries, leading to downstream tissue hypoxia, necrosis and pain.

Malarial parasites invade the red blood cells via mosquito bites and consume intracellular oxygen, triggering sickling and the selective removal of parasitized cells in the heterozygous carriers. Those individuals with the sickle cell trait therefore appear to be **more resistant to malaria.**

It has long been a puzzle why the fatal disease cystic fibrosis is so common in the Caucasian population. In the homozygous affected state, the thick mucus produced on the surface of the body's epithelial cells clogs the lungs and intestines. This is due to a fault in the chloride channels of the cell membrane. Because chloride ions cannot pass out of the cell, water is also prevented from exiting into the lungs or intestines, and the mucus normally present there is insufficiently hydrated.

This means that the body fluids are not as dilute as they should be. It has now been suggested that there may be a heterozygote advantage such that in the past a carrier of CF may have had more resistance to cholera, which led to death from dehydration due to rapid loss of water through diarrhoea. Perhaps the carrier had more resistance to water loss than the normal person.

Linkage and LOD scores

Two genes close together on the same chromosome are said to be linked. Before DNA testing it was often noticed that two separate characteristics could run in the same family such that the affected

Another example of linkage is between the dominant disease myotonic dystrophy and the secretor locus (i.e. positive or negative for a substance in the saliva) on chromosome 19. In a small nuclear family it might be that all the affected persons had the negative secretor allele, whereas those unaffected carried the positive secretor allele. The locus for secretor was so close to the locus for myotonic dystrophy that there was very little chance of a crossover of alleles at meiosis, so the two alleles remained linked.

There is an association between the amount of recombination and map units which determine the distance apart of two loci. One **centimorgan** is equivalent to a 1% recombination rate.

The example below uses **logarithms (logs) to the base 10** ($\log_{10}$). Large numbers are expressed as powers of 10, so for example $1000 = 10^3$; so $\log_{10}1000 = \mathbf{3}$. On the same principle, $\log_{10}100 = 2$, $\log_{10}10 = 1$ and $\log_{10}1 = 0$.

Where two numbers are ordinarily multiplied, the logs of each of those numbers are added (e.g. $M \times N = MN$; $\log(MN) = \log M + \log N$). Powers of numbers can be multiplied with logs (e.g. $\log_{10}4^7 = 7\log_{10}4$).

individuals also had the same allele for an unassociated physical characteristic.

Sufferers of the autosomal dominant nail patella syndrome (oddly shaped fingers, toenails and kneecaps) in a particular family were sometimes found to have the same allele from the gene that determines ABO blood groups, whereas the normal individuals had a different allele. The reason was that both nail patella syndrome and the genes for the ABO blood group are found close together on chromosome 9; they are linked. Although coding for completely separate characteristics, the alleles are passed on together as a haplotype.

With the advent of DNA testing which can distinguish different polymorphisms (often called **markers**), it becomes useful if it can be established whether the polymorphisms are linked to any genetic disorders.

If a marker allele can be shown to be linked to a disorder in a family, the disease can be **tracked** through that family even if we do not know the exact gene locus of the disease (see Chapter 7).

By using the amount of recombination between a disorder and a marker, the distance between the two loci can be estimated. This can be done empirically (i.e. by looking at existing cases in families and collecting data).

It is obviously important to know how far away a polymorphic marker is with respect to a disorder where the gene locus is not known, as the closer the marker, the more tightly linked they will be and the less will be the recombination.

This would mean that a close, tightly linked marker would more accurately track the disorder in a family, and the predictive value would increase.

A statistical method can be used which estimates not only the chance that a disease might be linked to a marker, but even the most likely recombination frequency between the disease and the marker. This involves calculating **LOD scores** (log of the **o**dds on linkage).

Taking the example of the family in Figure 4.6, we see that some members are affected with an autosomal dominant disease. A

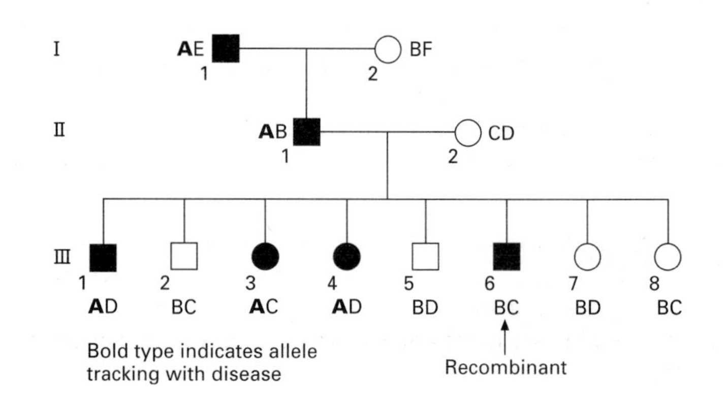

Figure 4.6 *Calculating LOD scores*

marker suspected to be linked to this disease has six alleles: A, B, C, D, E or F.

Chromosomes come in pairs, so there is always a choice of two alleles out of these six.

Notice that most of the affected members in this family carry allele A, suggesting that this allele is linked to the disease and is inherited along with the abnormal disease gene.

One affected member of the family has not inherited allele A; there has been a recombination between the marker locus and the disease locus. The number of recombinants in a family gives us an estimate of the recombination risk (here, one person/meiosis in eight).

In our family there is only one recombinant in eight affected people; if these loci were unlinked we would expect 50% recombination ($\theta = 0.5$), that is four recombinants out of eight.

However, we have to look at the family structure as a whole to calculate the LOD scores at different values of θ. This is because we have to take into account the fact that two loci could be linked by chance.

We deduce that allele A tracks along with the disease as affected members of three generations show this pattern (with the exception of the recombinant). This is called knowing the **phase**, that is, in II_1 allele A tracks with the disease and allele B tracks with the normal state. LOD scores become more difficult to calculate if some crucial family members have died and phase is not definitely known.

The LOD score for the family structure is called Z and the recombination fraction (rate) is represented by θ. In this family, seven people have not recombined $[(1-\theta)^7]$ and one person has recombined (θ). The chance of no linkage in eight people is 50% or $[(1/2)^8]$.

The ratio (or likelihood) of linkage to non-linkage is therefore:

$$Z = \log_{10}[(1-\theta)^7 \times \theta \text{ divided by } (1/2)^8]$$

$$Z = 7\log_{10}(1-\theta) + \log_{10}\theta + 8\log_{10}2$$

You now calculate Z for each of several recombination fractions (usually a range from q = 0 to q = 0.5):

θ	0	0.1	0.2	0.3	0.4	0.5
Z	−infinity	1.0874	1.0308	0.8009	0.4575	0

Usually there is said to be linkage if $Z > 3$, and no linkage if $Z < -2$.

Notice that in our example the family is not big enough to reach a score of 3; often the results of several families have to be combined to be statistically significant.

Figure 4.7 shows some examples of results that might be obtained after performing calculations of LOD scores for three different markers:

(A) Probable linkage at odds of 1000 : 1, no recombinants.
(B) Possible linkage at odds of 1000 : 1 but a recombination rate of 0.3.
(C) A lower likelihood of linkage (10 : 1) at a recombination rate of 0.3.

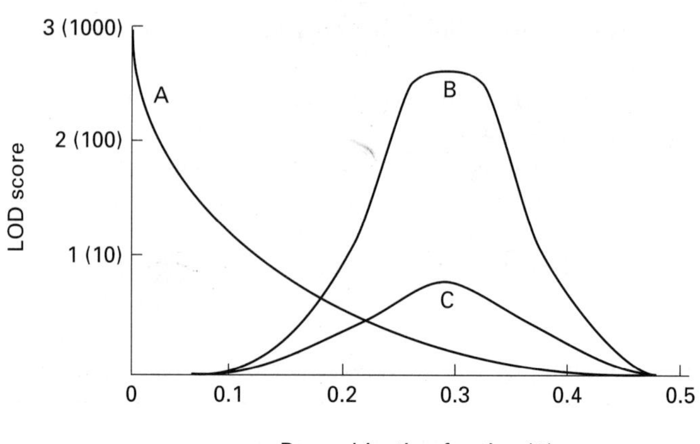

Figure 4.7 *Using LOD scores to estimate recombination risk*

A: LOD = 3 (1000:1 odds on linkage) with $\theta = 0$
B: LOD = 3 (1000:1 odds on linkage) with $\theta = 0.3$
C: LOD = 1 (10:1 odds on linkage) with $\theta = 0.3$

Bayes' theorem

Another very useful mathematical equation was named after its inventor, Reverend Bayes, in 1763.

We may wish to know the chances of someone being a carrier of a disease, but we must also take into account the chance of them not being a carrier. In a similar manner we can compare the chance that someone is affected with a disease versus the chance that they are not affected.

The advantage of Bayes' theorem is that it can **incorporate new information** contributing to the final risk.

We usually start with the initial estimated risk taken from pedigree information. This is called the **prior risk**. This risk can then be modified, making it more or less likely that someone is a carrier or is affected.

Many different factors may be taken into consideration; these are called the **conditional risks**. If there is more than one condition they can be multiplied together. Examples of such conditions include:

- numbers of normal children;
- age (with a late onset disease);
- enzyme levels;
- mutations already tested for by DNA.

An **index case** is the first affected individual in a family to be investigated for a particular disorder. A **consultand** is any individual attending a clinic for a **consultation** with a medical specialist, in this case a **consultant** clinical geneticist.

The consultand from the pedigree in Figure 4.8 is asking if she is a carrier of Duchenne muscular dystrophy. This is an X-linked disorder, and her sister has an affected son, and an affected brother, suggesting that she is a carrier. Look at Table 4.7.

The grandmother I_2 has one abnormal X chromosome and one normal chromosome. There is therefore a one in two chance that

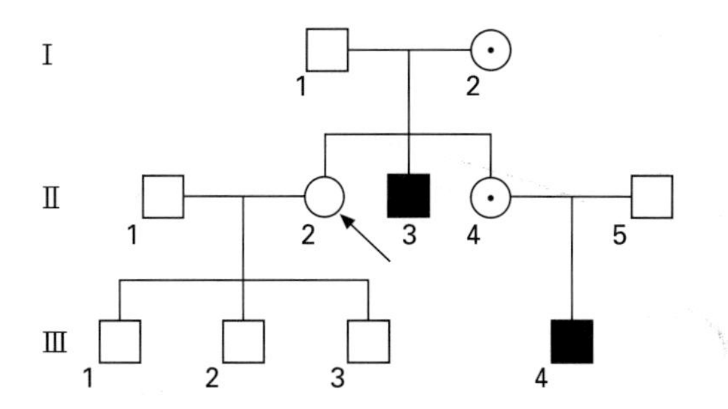

Figure 4.8 *Use of Bayes' theorem to calculate a DMD carrier risk*

Table 4.7 *Bayes' theorem used in Duchenne's muscular dystrophy*

	Carrier	*Not a carrier*
Prior risk	1/2	1/2
Conditional risk	1/2 × 1/2 × 1/2 = 1/8	1
Joint risk	1/16	1/2

she has passed the abnormal X to her daughter (prior risk of being a carrier = 1/2). The chance she has passed on the normal chromosome (i.e. not a carrier) is also 1/2.

The consultand II_2 has already had three normal sons. Common sense tells you that this makes her more likely not to be a carrier. If she **is a carrier,** the chance of her having a normal son is 1/2 (i.e. of passing on her normal X). As she has had three normal sons this would have happened three times; these chances are multiplied together giving a conditional risk of $(1/2)^3$ (i.e. 1/2 × 1/2 × 1/2 = 1/8).

If the consultand **is not a carrier** she can only have normal sons with a certainty of 1.

The prior risk is multiplied by the conditional risk to give a **joint risk.** This is the initial risk **modified** by the extra information we have incorporated.

These figures are now put into the Bayes' theorem to give a final carrier risk. This will be expressed in words as:

$$\frac{\textbf{The joint risk of being a carrier}}{\textbf{The joint risk of being a carrier + the joint risk of not being a carrier}}$$

Putting in the above figures, we get:

$$\frac{1/16}{1/16 + 1/2} = \frac{1/16}{9/16} = 1/9$$

Her risk has therefore been reduced from 1/2 to 1/9.

In the next example (see Table 4.8) a phenotypically normal individual is wishing to know their carrier risk for the autosomal recessive disease cystic fibrosis. The carrier risk in the Northern European population is around 1/20. The prior risk of that person not being a carrier is therefore 19/20 (1−1/20).

We could also incorporate the results of testing for the enzyme creatine kinase, which is present at a higher level in female carriers. The reading may modify the risk towards or away from being a carrier, depending on whether the level is high or low respectively.

Table 4.8 *Bayes' theorem used in cystic fibrosis*

	Carrier	*Not a carrier*
Prior risk	1/20 (0.05)	19/20 (0.95)
Conditional risk	10% (0.1)	1
Joint risk	0.005 (0.05×0.1)	0.95 (0.95×1)

A DNA laboratory tests their DNA for a range of CF mutations, known to account for around 90% of all CF mutations in the local population, and finds them negative (i.e. normal) for those mutations.

The only way he or she can still be a carrier is if he or she carries a mutation present in the 10% not tested for; this is the conditional risk. Someone who is not a carrier will by definition not carry any CF mutations – with a certainty of 1 (see page 81).

Applying the formulation as for the first example, the final carrier risk will be:

$$\frac{0.005}{0.005 + 0.95} = \frac{0.005}{0.955} = 1/191$$

The carrier risk for this individual after a negative DNA test has been reduced from 1/20 to 1/191.

Summary

Classical genetics includes Mendel's laws and the major Mendelian inheritance patterns, and is the foundation upon which the prediction of familial single gene disorders is based.

Risks formerly calculated solely from pedigrees can now be augmented by batteries of medical tests, and statistical calculations have been developed in order to incorporate these results into a more accurate risk estimate.

Although thousands of single gene disorders are already characterized, there still remains a vast range of diseases, such as autoimmune disorders and heart defects, for which the genetic basis is strongly modified by environmental factors.

Suggested further reading

Harper, P.S. (1998). *Practical Genetic Counselling*. 5th Edn. Butterworth-Heinemann.

Read, A.P. and Newton, V.E. (1997). Waardenburg syndrome. *Journal of Medical Genetics*, **34**, 656–665.

Sinnott, E., Dunn, L. and Dobzhansky, T. (1958) Mendel's law of segregation, in *Principles of Genetics*, 5th Edn. McGraw-Hill.

Vines, G. (1998). Inheritance. *Science*, November, 27–30.

Young, I.D. (1991). *Introduction to Risk Calculation in Genetic Counselling*. Oxford University Press.

Self-assessment questions

1. The recombination fraction for two genes is 0.1. What does this tell you about these genes? Explain your reasoning.
2. Name the three major types of inheritance patterns and give an example of a disorder for each one.
3. Albinism (a lack of pigment in the skin and eyes) is an autosomal recessive disease. A man has a sister who has this disorder. What is the chance that he is a carrier of albinism?
4. A male and female both affected with achondroplasia decide to marry and have children. Given that the homozygous dominant form is lethal prenatally, what proportion of their children will be of normal height?
5. An adult male patient has increasing loss of vision, which started at the outer edges of his eyes and is progressing inwards giving him tunnel vision. After examining the family pedigree, the clinical geneticist tells him that he has X-linked retinitis pigmentosa. This man has three daughters and one son (all over 16 years old). How would the clinical geneticist counsel the man and his partner with respect to:
   (a) their carrier risk?
   (b) their children's carrier risk?
6. A man who has azoospermia (produces no mature sperm) wishes to undergo infertility treatment by intracytoplasmic sperm injection (ICSI), surgically using one of his mature sperm to fertilize an egg. It is found by molecular analysis that he has a deletion at the azoospermia locus on his Y chromosome. What are the prospects for his male children?
7. An adult male had a paternal grandmother who died of Huntington disease (HD). His father is 65 years old and has not yet developed the disease. Using Bayes' theorem, and given that by 65 years of age, 85% of people carrying HD would have developed the disease, what is the risk that his father is carrying the gene? What is the risk of the son having HD?

Key Concepts and Facts

Mendel's Laws

- Mendel's first law (the principle of segregation) states that during meiosis each of a pair of alleles in a diploid cell segregates into different haploid cells.

- Mendel's second law (the principle of independent assortment) states that genes on different chromosomes segregate independently of one another at meiosis.

Linkage and Recombination

- The more closely linked genes are on a particular chromosome, the less chance there is of a crossover and hence genetic recombination. The maximum possible recombination rate for unlinked genes is 50%.

Mendelian Inheritance Patterns

- There are three major modes of inheritance: autosomal dominant, autosomal recessive and X-linked recessive. There are two less common patterns: X-linked dominant and Y-linked.

Complications Disrupting Mendelian Inheritance Patterns

- Within each mode of inheritance, such factors as reduced penetrance or variable expressivity may mask the pattern, or there may be more global problems such as heterogeneity, imprinting, new mutations, or germinal mosaicism. Mitochondrial inheritance is maternal.

Non-Mendelian Inheritance

- This may arise from the combined action of polygenes, or may be due to a varying contribution from the environment, such that a susceptibility gene may produce different phenotypes depending on exposure to a particular environment.

Population Genetics

- Alleles of genes in balance in a population are distributed according to the Hardy–Weinberg equation. Various factors may disturb this equilibrium.

Statistics Used in Genetics

- LOD scores are a means of analysing the extent to which genes are linked. A score of >3 indicates strong linkage. Bayes' theorem is a flexible equation which weighs up the probability that an individual may be a carrier or affected by a disease.

Chapter 5
Cytogenetics

Learning objectives

After studying this chapter you should confidently be able to:

Explain what is meant by a balanced and an unbalanced chromosome complement.

Understand why some chromosome abnormalities have a direct phenotypic effect while others have reproductive implications.

List the cell types commonly used for chromosome preparation.

List the common types of structural chromosome abnormalities.

Name some common chromosomal syndromes and the chromosome defects responsible.

Describe the effects of the common sex chromosome syndromes and compare them with the general features common to autosomal syndromes.

Cytogenetics is the visual study of chromosomes at microscopic level. As with any branch of genetics, the general principles which apply to the behaviour of chromosomes of other organisms may be applied to humans. The chief area of interest in human cytogenetics is the identification of those situations in which an abnormality of the chromosome complement is responsible for medical problems.

With few exceptions, **chromosome analysis is performed on mitotic** (rather than meiotic) **chromosomes**, using a conventional light microscope equipped with binocular eyepieces and a high-quality oil-immersion objective lens, at a magnification of about $\times 1000$.

Analysis of chromosomes in **meiosis** is rarely undertaken. For technical reasons, and because of the timing of the cell cycle, obtaining analysable chromosome preparations from the female oocytes is almost impossible. Male spermatocyte preparations are possible, but the patient has to undergo a testicular biopsy, and in reality this approach is used only in specialized research.

The karyotype

'**Karyotype**' means 'chromosome complement'. It can be used in the sense of the chromosome complement of a species, a single organism, or even an individual cell. The word is also applied to a picture in which the chromosomes are cut and pasted into matching pairs (see Figure 1.8).

The normal human somatic cell has a diploid complement of 46 chromosomes with two copies, or **disomy**, of each chromosome, one of paternal and one of maternal origin. The chromosomes include the two sex chromosomes, XX female, or XY male respectively, and 44 **autosomes** (chromosomes not directly involved in sexual differentiation) numbered approximately in descending

Ploidy refers to the number of chromosome sets in a karyotype. The single set of chromosomes in the sperm or the egg is known as **haploid**. The two haploid gametes, each with 23 chromosomes, fuse to give a diploid zygote with 46 chromosomes. Other levels of ploidy are occasionally seen. **Triploidy**, three complete chromosome sets, occurs most commonly when an ovum is fertilized by two sperm, the result being an abnormal pregnancy that usually miscarries. **Tetraploidy**, four chromosome sets, occurs when the nucleus and cytoplasm of a diploid cell fail to divide following normal replication and division of the chromosomes. It is seen sometimes in tissue culture, in the placenta or in tumours.

Down syndrome, the result of trisomy for chromosome 21, is the most familiar chromosome syndrome and is a typical example of recognizable specific facial dysmorphism, with the characteristic small nose, flat profile both to the face and the back of the head, and distinctive shape to the eyes resulting from epicanthic folds. The ears are small and low-set. The stature is short. Mental handicap is universal, and heart defects common.

order of size from 1 to 22. Deviation from this normal balanced complement results in clinical abnormality. It is extremely important to understand that, in most cases where an abnormal phenotype is associated with an abnormal karyotype, the phenotype is usually attributable to extra or missing chromosome material, that is an imbalance, or dosage effect, of normal genes rather than to particular defective genes.

This is an important difference between cytogenetic syndromes and single gene disorders (see Chapters 4 and 7). The commonest types of unbalanced karyotype involve **monosomy** (one copy) or **trisomy** (three copies) for all or part of a chromosome resulting from a numerical or structural abnormality. **Nullisomy** (no copies) and **tetrasomy** (four copies) are rarely encountered. There are also other structural abnormalities of the chromosomes which, while balanced and therefore compatible with a normal phenotype, generate unbalanced forms at the meiotic division of gametogenesis, resulting in an abnormal phenotype in the following generation.

With a few exceptions, gain or loss (trisomy or monosomy) of a specific region of the genome is responsible for the expression of a particular spectrum of clinical abnormalities and thus there are many identifiable chromosomal syndromes attributable to specific gain or loss. Most chromosome syndromes include mental handicap, and many have general features in common such as growth retardation, skeletal malformations, heart and other organ defects, and midline defects such as cleft palate. Often a syndrome is associated with distinctive dysmorphic facial features peculiar to that syndrome.

There is a limit to the extent of imbalance compatible with life, and many conceptions with an unbalanced karyotype abort spontaneously. There are some chromosome regions, probably containing vital developmental genes, for which imbalance cannot be tolerated.

Chromosome preparation

Individual chromosomes are not distinguishable in the interphase nucleus, but become visible only as they condense immediately preceding mitotic or meiotic division.

In the majority of somatic tissues there is insufficient mitotic activity to retrieve chromosome preparations directly from the tissue, and it is necessary to resort to tissue culture to produce the necessary dividing cells. Two exceptions are the chorion of the placenta, where the cytotrophoblast cells display spontaneous activity (see Chapter 9 on Prenatal diagnosis), and the bone marrow (see Chapter 8 on Cancer genetics). A variety of tissue culture methods are appropriate to different tissues (see Table 5.1).

Cells are grown by conventional tissue culture methods in sterile liquid medium. When sufficient mitotic activity is occurring, the

Table 5.1 *Tissues cultured for cytogenetic studies*

Tissue	*Type of culture*	*Culture duration*	*Reason for referral*
Blood (lymphocytes)	Suspension	2–3 days	Wide range of postnatal reasons
Amniotic fluid	Monolayer	7–14 days	Prenatal diagnosis
Chorion	Direct	0–24 hours	Prenatal diagnosis
	Monolayer	7–14 days	
Skin/organ	Monolayer	7–21 days	Abortion/post mortem, investigation of mosaicism
Bone marrow	Suspension	0–72 hours	Leukaemia
Tumour	Suspension or attached	Varies	Solid tumours

cells are arrested at metaphase of mitosis by the addition to the culture of colchicine or its synthetic derivative colcemid. The effect is to destroy the mitotic spindle fibres so that the chromosomes, randomly distributed within the cell rather than on the metaphase plate, are unable to proceed to anaphase. Depending on the tissue involved, the exposure to the arresting agent may be from a few minutes to a few hours. Following this the cells are exposed for some minutes to a hypotonic salt solution, for example potassium chloride or sodium citrate, in order to swell the cells and facilitate separation of the chromosomes, and then fixed in a mixture of methanol and acetic acid. The chromosome preparation is made by placing a drop of concentrated fixed cell suspension on to a microscope slide and simply allowing the fixative to evaporate.

The ideal is to produce chromosomes that are long and well defined with little overlapping. The quality of the preparation varies depending on the cell type, with lymphocyte culture generally giving the best results. Amniotic fluid cells are most commonly used for prenatal diagnosis (see Chapter 9), while lymphocytes are the cells of choice for most postnatal chromosome studies. The latter are easily obtained from a small blood sample, they are easily grown, and they provide a high quality result.

Lymphocytes are a type of white blood cell concerned with immune response. A particular subtype, the T cell, normally does not divide actively, but can be stimulated to do so by adding a substance called phytohaemagglutinin to the culture medium. In the standard blood culture, the lymphocytes are not separated from the blood sample, but a few drops of whole blood are simply added to the culture medium. Most of the unwanted components of the blood are destroyed when the culture is finally subjected to hypotonic treatment and fixation. **Lymphocyte cultures** are often synchronized by temporarily blocking the passage of cells through the DNA synthesis phase of the cell cycle by the addition of a high concentration of thymidine or an antimetabolite drug such as methotrexate. Following release of the S-phase block it is possible to harvest a high yield of cells in late prophase to early metaphase where the chromosomes are relatively uncontracted.

Chromosome analysis

When undertaking chromosome analysis, the total chromosome number, the morphology of each chromosome with respect to its length and centromere position, and the staining pattern are all taken into account.

For most routine investigations, analysis makes use of the **G-banding** technique, a method in which a cross-banding pattern unique to each chromosome is achieved by a variety of treatments. The most widely used and simplest method entails dipping the slide in a solution of the enzyme trypsin before staining the slide in Giemsa or Leishman's stain. Using this method, the trained cytogeneticist is able to identify each pair of chromosomes and

Table 5.2 *Staining methods for chromosome analysis*

Method	*Displays*	*Use*
G-banding	Unique pattern for each chromosome	Routine analysis
Block (or solid) staining	Length and centromere position only	Some aspects of morphology, aberrations
C-banding	Constitutive heterochromatin	Defining centromeres, markers
R-banding	Late replicating X	Sex chromosome abnormalities
Q-banding	Y chromosome, some other regions of heterochromatin	Y chromosome abnormalities or mosaicism
Silver staining	Nucleolus organizer regions	Defining acrocentrics and markers
In situ hybridization	Specific region of probe homology	Microdeletions (see Chapter 6)

The **centromere** is the site on the chromosome which controls division of the two chromatids into daughter chromosomes at mitosis. It is visible as a distinct constriction, the **primary constriction**, in the metaphase chromosome and, as it occupies a consistent location in the chromosome, it is an important factor in chromosome identification.

spot abnormalities of the banding pattern. Various other staining methods, outlined in Table 5.2, may be used to highlight specific regions of the karyotype, such as heterochromatin and nucleolar organizer regions occurring at particular chromosomal locations.

Chromosome preparations are also analysed by *in situ* hybridization, a specialized technique dealt with in detail in Chapter 6.

Analysis at the microscope entails firstly scanning the slide using a low power objective lens in order to locate suitable chromosome spreads, then transferring to a high power image, counting the number of chromosomes and, lastly, visually assigning the chromosomes to their pairs. In some cases, microscopic analysis is supplemented by the production of a permanent karyotype; this may be produced by taking a photograph of the chromosome spread and cutting and pasting the chromosomes into pairs. Alternatively, there are now computerized image-processing systems with which the cytogeneticist can manipulate digitized images of the chromosomes to produce a karyotype of photographic quality (see Figure 1.8). A minimum of four or five cells needs to be analysed in order to be confident of excluding an abnormality; in some circumstances in which there is the possibility of more than one cell line with different chromosome complements, a phenomenon known as mosaicism (explained later in this chapter), many more cells have to be examined.

The number of bands visible along a chromosome depends on the extent of its condensation (or contraction). Different numbers of bands tend to be seen in typical metaphase cells of different tissues, probably the result of cell cycling characteristics and basic cell morphology. Because of these differences, certain cell types tend to yield more highly condensed chromosomes.

The limit of resolution of chromosome analysis is the detection of gain or loss of a single band. A good quality lymphocyte preparation contains 1100–1600 dark and pale bands per karyotype; the smallest bands contain approximately three to five million base pairs, and so potentially many hundreds of genes may be lost or duplicated in even the smallest unbalanced structural chromosome abnormality. Chromosomes from amniotic fluid cell cultures, the usual choice for prenatal diagnosis, usually yield 700–1100 bands, while chromosomes of malignant tumour tissues, frequently poorly defined and difficult to identify with the same level of accuracy as chromosomes from other tissues, may have as few as 100–400.

Reliable detection of subtle abnormalities of the banding pattern also depends on the clinical information accompanying the request for analysis. It is much easier to detect, for example, a small deletion if the clinical features of the patient suggest a specific deletion syndrome.

Chromosome abnormalities

Autosomal aneuploidies

The common viable aneuploidies are trisomy 21 (Down syndrome), trisomy 18 (Edwards syndrome) and trisomy 13 (Patau syndrome). Trisomy for almost all the other chromosomes has been demonstrated in tissue from spontaneous abortions. An example of standard nomenclature for trisomy would be 47,XX,+21, female Down syndrome.

Currently in the UK, about 60% of trisomy 21 pregnancies are diagnosed prenatally, while most of the remainder are detected at birth where the baby will display characteristic features familiar to the paediatrician. Confirmation of a trisomy 21, 18 or 13 can be given to the paediatrician about 48 hours after receipt of the blood sample in the laboratory. Except for rare cases of mosaicism (see later), no other autosomal aneuploidies are found in the liveborn, although they contribute significantly to spontaneous abortions.

Structural abnormalities

There are many factors which may cause a break or a discontinuity in a chromosome. Sometimes, under the influence of DNA repair mechanisms or simply during replication, broken ends may rejoin incorrectly to generate structurally abnormal chromosomes. Some karyotypes with a structural abnormality are **balanced** and may occur in people who have no direct health problems; those that are **unbalanced**, with a partial monosomy or partial trisomy, have serious clinical implications.

Some of the commonest structural abnormalities involve two breakpoints at which the four broken ends rejoin incorrectly, as illustrated in Figure 5.1 where (a) generates a pericentric inversion, (b) a paracentric inversion, (c) a ring chromosome, (d) an interstitial deletion and (e) a reciprocal translocation.

Deletions

A deletion involves loss of material from a chromosome. Two types of deletion are identifiable: the **terminal deletion**, a single break resulting in loss of the end of a chromosome, and the **interstitial deletion**, where two breaks result in a segment being lost from within a chromosome (Figure 5.1(d)). All deletions result in an unbalanced karyotype, with partial monosomy, and therefore the result is almost inevitably one of serious clinical effect. A number of clinical syndromes have been ascribed to specific deletions.

Autosome is a convenient general term applied **to any chromosome that is not a sex chromosome**. **Aneuploidy** refers to any chromosome complement not exactly haploid, diploid, triploid, tetraploid, and so on. In human cytogenetics, aneuploidy almost always refers to deviation from the diploid chromosome number.

Edwards syndrome occurs in 1 in 3000 livebirths. Birth weight is low. The chin is small, the ears malformed and there is a distinctive shape to the back of the head. The hands are clenched with overlapping index and fifth fingers. The feet are distinctive, with a shape usually described as 'rocker-bottom'. Only a small proportion survive infancy, and these children are profoundly delayed.

Patau syndrome occurs in 1 in 5000 livebirths and includes cleft lip and palate, small eyes and polydactyly. Congenital heart defects are common, and survival beyond infancy is rare.

About 20% of recognizable pregnancies abort spontaneously, and of these it is estimated that up to half may have a chromosome abnormality. Amongst those with an abnormal karyotype, about 50% have an autosomal trisomy, 20% have 45,X (Turner syndrome), and 16% have triploidy. More than a third of the trisomic abortions have trisomy 16, an abnormality incompatible with survival to birth.

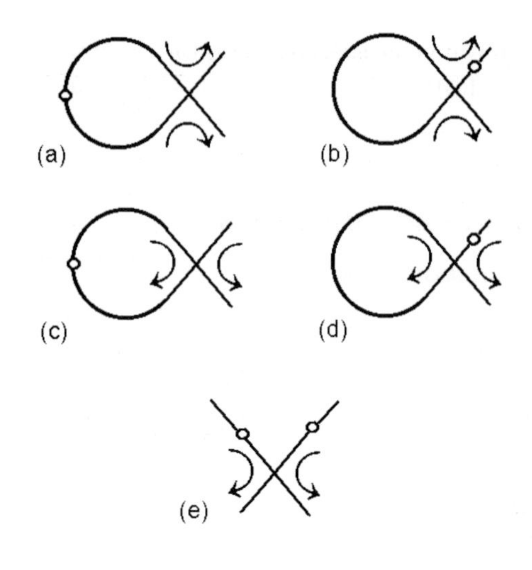

Figure 5.1 *Some common chromosome rearrangements arise by reciprocal exchange at two breakpoints on the same chromosome (a–d) or different chromosomes (e): (a) pericentric inversion, (b) paracentric inversion, (c) ring chromosome, (d) interstitial deletion and (e) reciprocal translocation*

The ISCN nomenclature easily distinguishes between terminal and interstitial deletions: for example, a terminal deletion of the long arm of chromosome 10 in a male might be 46,XY,del(10)(q26), while an interstitial deletion might be 46,XY,del(10)(q24q26).

Deletions range in size from large ones readily detected in standard G-banded chromosome preparations, to small ones at the limit of microscopic resolution, through to those very small ones requiring *in situ* hybridization to detect. Beyond this, there are deletions within a gene only detectable by molecular methods: microscopically detectable interstitial chromosome deletions are simply the opposite end of the same spectrum.

New rare **deletion syndromes** continue to emerge, more than 30 years after the first description of cri du chat syndrome (see Table 5.3 and Figure 5.2). The identification of these often tiny deletions needs a combination of excellent technical and analytical skills on the part of the cytogeneticist and the critical expertise of the specialist clinician. Recent examples, only beginning to be recognized, include terminal deletion of 1p36 (the tip of the short arm of chromosome 1) and of 2q37 (the tip of the long arm of chromosome 2).

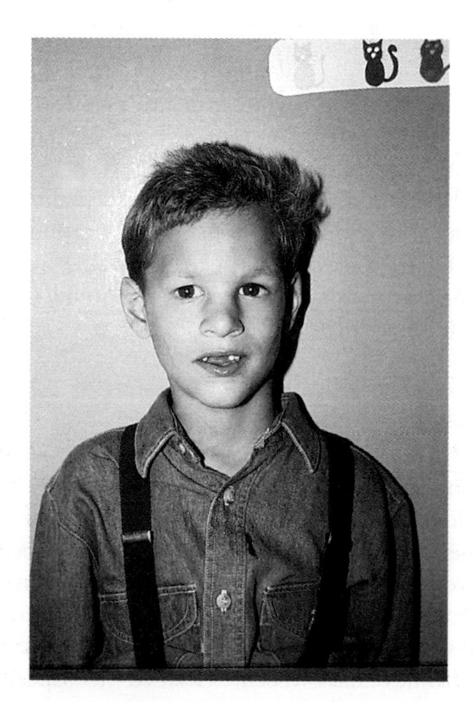

Figure 5.2 *Cri du chat syndrome*

Table 5.3 *Some autosomal deletion syndromes*

Syndrome	*Deletion*	*Type*
Wolf–Hirschhorn syndrome	4p36	Terminal
Cri du chat syndrome	5p15	Terminal
Williams syndrome	7q11.2	Interstitial microdeletion, detectable only by FISH
Langer–Giedion syndrome	8q24	Interstitial
9p – syndrome	9p13	Terminal
Jacobsen syndrome	11q23	Terminal
Retinoblastoma	13q14	Interstitial
Prader–Willi syndrome	15q11.2	Interstitial
Angelman syndrome	15q11.2	Interstitial
ATR-16 syndrome (α-thalassaemia with mental retardation)	16p13	Terminal
Miller–Dieker syndrome	17p13	Terminal
Smith–Magenis syndrome	17p11	Interstitial
18p – syndrome	18p11	Terminal
18q – syndrome	18q21	Terminal
DiGeorge/Shprintzen syndrome	22q11.2	Interstitial, rarely detectable without FISH

Isochromosomes

An isochromosome is a chromosome composed of **two copies of same arm.** In other words, it is a mirror image about the centromere. A mechanism of origin sometimes quoted is that of 'transverse division of the centromere'. Although in the light of current knowledge of centromere structure this is probably not strictly correct, it is a useful way of thinking of the composition of this type of structural abnormality.

As far as the occurrence of isochromosomes in the karyotype is concerned, it is convenient to place them into two groups. Firstly, there is the isochromosome which **replaces a normal chromosome.** In most chromosomes such a situation would result in an extreme and inviable imbalance, with monosomy for one complete chromosome arm and trisomy for the other arm of the same chromosome. However, familiar examples include the isochromosome of the long arm of the X found in about 10% of females with Turner syndrome (see the section on sex chromosome abnormalities), where the chromosome number is 46, there is one normal X and one so-called 'iso-Xq'. An isochromosome of the long arm of chromosome 21 is found in about 1–2% of cases of Down syndrome. Here the chromosome number is 46, rather than the expected 47, with the trisomy attributable to the presence of a normal 21, plus two copies of the long arm of 21 in the form of an isochromosome.

If a terminal deletion were as simple as the loss of the end of a chromosome, there would be no telomere to stabilize the tip of the deleted chromosome. It is widely accepted that any **stable chromosome arm must have a telomere**, including an arm which under the microscope appears to have a straightforward terminal deletion. Where a deleted chromosome arm gets its telomere is not known for certain but the possibilities are as follows:

- There is activation of an ancestral interstitial telomere located at the breakpoint.
- A new telomere is manufactured by normal cellular processes.
- The rearrangement actually involves a second breakpoint close to a telomere; in other words, it is a cryptic interstitial deletion or a translocation product.

Some ESACs (extra small accessory chromosomes), explained later in the chapter, may be isochromosomes.

A complication associated with some **isochromosomes** is that they tend to occur in **mosaic form**, for example the iso-Xq (in some Turner syndrome females) and the iso-12p (Pallister–Killian syndrome). This is possibly related to the involvement of the centromere in the structural changes: alteration of the normal structure of a centromere may interfere with its function in controlling separation of the chromatids at anaphase. Some isochromosomes can indeed be shown to possess two centromeres.

Visually, it may not be possible to distinguish between an isochromosome and a centromere to centromere **translocation involving two homologous chromosomes**. Some of these structural changes are discussed in the section on Robertsonian translocations.

The ISCN nomenclature easily distinguishes between paracentric and pericentric inversions: for example, a paracentric inversion of the long arm of chromosome 10 in a male might be 46,XY,inv(10)(q22q26), while a pericentric inversion which has a short arm (p) and a long arm (q) breakpoint might be 46,XY,inv(10)(p12q26).

In the second group, the isochromosome is present as a **supernumerary chromosome**. Examples of this include the 'iso-12p' found in Pallister–Killian syndrome, and the 'iso-18p' which results in tetrasomy for the short arm of chromosome 18 and is associated with a defined syndrome of dysmorphism and mental handicap.

Inversions

An inversion is the **reversal of a segment of a chromosome**. It requires two breakpoints and, as it is balanced, it is generally compatible with normal development. The carrier of an inversion may be at risk of producing gametes with an unbalanced karyotype and, depending on the structure of the inversion (that is, the location of the breakpoints), the result may be reduced fertility, miscarriage or the live birth of a handicapped child.

It is important to differentiate the two types of inversion, **paracentric** and **pericentric**. In a paracentric inversion the reversed segment is within one chromosome arm (Figure 5.1(b)), while in a pericentric inversion the reversed segment includes the centromere (Figure 5.1(a)). The consequences of meiotic recombination and the reproductive risks are very different for these two rearrangements.

At meiosis, inverted chromosome segments adopt a loop formation for full pairing. Figure 5.3 illustrates just one chromatid of each chromosome: a normal one ABCDE, pairing with an inverted one ADCBE. If there is a meiotic crossover somewhere between B and D, then the two recombinants will be ABCDA and EDCBE, which are clearly abnormal and unbalanced.

In a paracentric inversion, the centromere is at A, so the recombinant chromatids are either dicentric or acentric, and therefore unstable. However, in a pericentric inversion with the centromere at C, the products are stable. Thus paracentric inversions generally have a low reproductive risk, while a pericentric inversion may have a high risk, especially if the inverted segment BCD is large, when the chance of recombination within the loop may be high. If the segments at A and E (that is, the ends of the chromosomes beyond the inversion) are short, the amount of imbalance will be small so that the unbalanced forms are more likely to result in viable conceptions. The cytogeneticist has to think

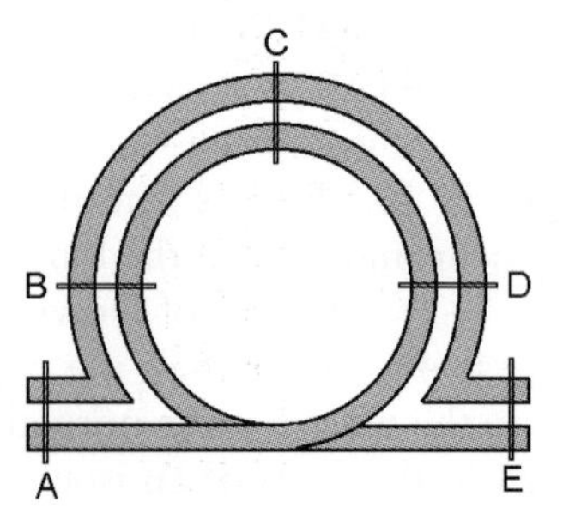

Figure 5.3 *Pairing of an inverted chromosome segment ADCBE with its normal homologue ABCDE. Crossing over within the loop results in recombinants ABCDA and EBCDE*

carefully about the breakpoints involved in a pericentric inversion in order to make a meaningful assessment of the reproductive risk. The maximum theoretical risk of imbalance is 50%.

Pericentric inversion of a small region of inert heterochromatin located around the centromere of chromosome 9 occurs in about 1% of the human population. It is completely benign. Other rarer small pericentric inversions are also considered to be **benign variants**, including those involving chromosomes 1, 2, 10 and 16.

Translocations

There are basically two types of translocation: Robertsonian and reciprocal. In the **Robertsonian translocation**, two acrocentric chromosomes, which are effectively V-shaped, join at the centromere into a single X-shaped structure. As two centromeres fuse into one, the balanced translocation karyotype has only 45 chromosomes.

The human acrocentric chromosomes 13, 14, 15, 21 and 22, which become involved in Robertsonian translocations, each have a very small short arm with no vital unique genes although, as explained already, they do have nucleolar organizers. The short arms are lost in the fusion that creates the translocation, but the imbalance has no clinical effect.

When pairing at meiosis, instead of the normal bivalent formation, there are three partially homologous components forming a **trivalent** (Figure 5.4). It is possible for the translocated chromosome to segregate into the same gamete as one of the normal chromosomes, resulting in the generation of genetically unbalanced gametes.

About one person in 1000 carries a Robertsonian translocation, more than half of which involve chromosomes 13 and 14. The majority of the remainder involve 14 and 21. As these translocations are among the commonest structural chromosome rearrangements found, the reproductive risks for them are well established. Carrying a 14;21 translocation confers a risk of a child with Down syndrome, and in a family with more than one affected child a 14;21 translocation is more often than not the culprit. Some carriers have a history of miscarriage; possibly some of their pregnancies had the non-viable trisomy 14. Naturally, the carrier of a Robertsonian translocation would be offered the opportunity of prenatal diagnosis by amniocentesis or chorion biopsy.

Statistically a woman carrier of a 14;21 translocation has approximately a 10% risk of having a liveborn Down syndrome child. If she is a known carrier and has amniocentesis at 16 weeks gestation, the chance of the pregnancy being affected is more like 15%. The difference in these two figures reflects the higher rate of loss through late spontaneous abortion of trisomic pregnancies. Curiously, in contrast, the male balanced carrier rarely fathers a trisomic child. Occasionally a child with trisomy 13 has been born to a 13;14 carrier: here the risk of trisomy is generally quoted as being about 1%.

A **reciprocal translocation** (see Figure 5.1(e) and Figure 5.5) arises when a two-way exchange of chromosome segments takes place, between two chromosomes, resulting in two **derivative** chromosomes. The exchanged portions are the **translocated segments**, while the remainder of each derivative chromosome is often called the **centric segment**. The reciprocal translocation is balanced, and therefore generally occurs in normal healthy people. The chromosome rearrangement is most often discovered when the carrier seeks

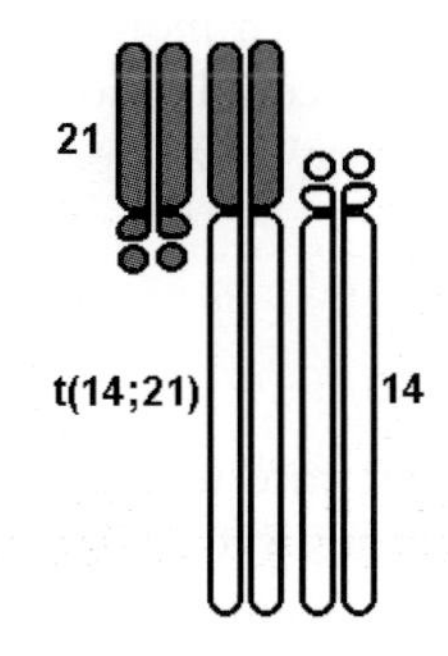

Figure 5.4 *Meiotic pairing in a Robertsonian 14;21 translocation heterozygote results in a trivalent configuration. If the translocation chromosome and the chromosome 21 both go into the gamete, the end result will be a Down syndrome pregnancy*

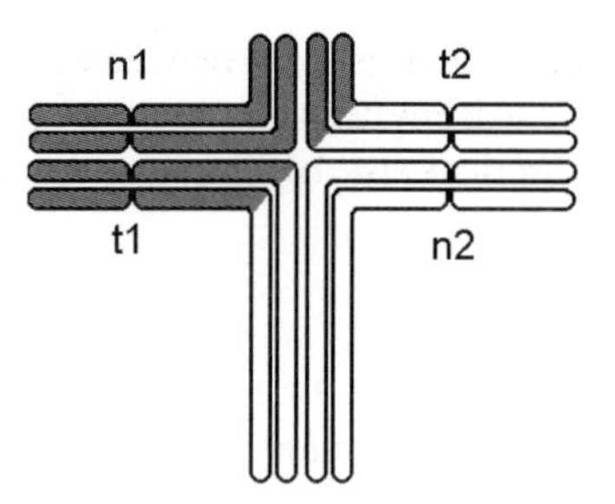

Figure 5.5 *Meiotic pairing in a reciprocal translocation heterozygote results in a quadrivalent formation. Contribution of **alternate** chromosomes n1 and n2 to the gamete is normal, and of t1 and t2 is balanced, but contribution of **adjacent** chromosomes t1 and n2 is one example of an unbalanced gamete*

> ISCN nomenclature for the male carrier of the reciprocal translocation shown in Figure 1.8, involving chromosomes 1 and 13, would be 46,XY,t(1;13)(q32;q33). A female pregnancy with an unbalanced karyotype having inherited the normal paternal 1 and derivative 13 would be 46,XX,der(13)t (1;13)(q32;q33) pat. This would be an example of adjacent segregation.

medical advice following miscarriages, or when a handicapped child is born, having inherited an unbalanced form of the translocation.

At meiosis, pairing and recombination takes place as normal, but there are homologous segments on four chromosomes, two normal and two derivative, which therefore come together as a **quadrivalent** (Figure 5.5). When the four chromosomes part company at anaphase of meiosis 1, they can enter the two daughter cells in almost any combination. There is one balanced (t1 segregates with t2) and one normal (n1 segregates with n2) combination, and from Figure 5.5 it may be seen that the **alternate** chromosomes segregate together. Although there are many other forms that are unbalanced, a gamete with extreme imbalance is unlikely to be viable, and for most translocations there are only one or two unbalanced types that can form recognizable pregnancies. These are usually where the two **adjacent non-homologous** chromosomes segregate together, that is n1 with t2, or n2 with t1.

> Occasionally a balanced translocation is found in someone with mental handicap or some kind of congenital or heritable physical problem or disease. Although this may sometimes be coincidence, there is evidence in some cases of disruption of a gene at one of the breakpoints, see position effect (Chapter 3). For example, the location of the gene for neurofibromatosis, NF1, was eventually shown to be on chromosome 17 after an affected person was shown to have a translocation involving that chromosome. In one family, adult polycystic kidney disease (APKD) was shown to segregate with a translocation involving chromosome 16. Duchenne's muscular dystrophy rarely affects girls, and some of those who do

Ring chromosomes

A ring chromosome is another manifestation of the incorrect rejoining of two breakpoints, one on each arm of a chromosome, such that a **circular chromosome containing the centromere** is formed (Figure 5.1(c)). Ring chromosomes are inevitably unbalanced since the tips of each arm distal to the breakpoints are lost; they are usually associated with partial monosomy and, like other unbalanced chromosome abnormalities, they have serious clinical effects. A further problem with a ring chromosome is that it may be somewhat unstable, with double-sized dicentric structures forming at replication; at mitosis the chromosome complement of the daughter cells may become even more unbalanced and eventually non-viable. Because normal cell division is compromised, growth retardation and failure to thrive may be noted in patients who have a ring chromosome.

Duplications

The duplication is another rare unbalanced type of chromosome abnormality and is almost never familial. It has the structure ABCDE<u>CDE</u>FGH, with a segment of chromosome repeated in tandem. Occasionally duplications may be inverted, that is ABCDEEDCFGH. As with other types of unbalanced abnormality, the patient is usually identified because of mental handicap or physical dysmorphic features.

have the disease have been found to have a translocation involving the critical locus on the X chromosome. So there are

There have been occasional cases of a clinically normal carrier of a ring chromosome. In such cases the breakpoints are so close to the ends of the chromosome that there is no significant loss of important genetic information. The carrier is generally identified because of reproductive problems, since there is a high probability of abnormal chromosomes being generated by meiotic recombination between the ring chromosome and its normal partner.

ESACs

An **extra supernumerary accessory chromosome**, or ESAC, is a structurally abnormal chromosome occurring in a karyotype in addition to 46 normal chromosomes. ESACs are mostly small, and their composition may be difficult to determine, requiring a battery of methods for their identification, for example *in situ* hybridization. The origin of such a chromosome, sometimes referred to as a **marker**, is variable: some originate as translocation products, others are tiny rings, isochromosomes or dicentric structures. The karyotype in which it occurs is always unbalanced, and the clinical effect depends on the composition of the ESAC. Some tiny accessory chromosomes which have effectively only a centromere and contain no important genetic information are benign, while others result in serious mental handicap or physical malformation. The identification of an ESAC at prenatal diagnosis poses a difficult dilemma, since the clinical effect is often impossible to predict.

Some rare structural chromosome abnormalities are more complex than those discussed in this chapter. They include balanced and unbalanced rearrangements which involve three or more breakpoints. Those which are balanced include three- or four-way translocations, insertions of interstitial chromosome segments in incorrect locations, and composite translocation/ inversions. They are often associated with a high reproductive risk owing to the much larger number of possible unbalanced forms, or infertility because of the difficulty of achieving complex patterns of pairing at meiosis.

Mosaicism

Mosaicism is the term applied to the situation where **two or more chromosomally different cell lines occur in a single person.** Co-existence of a normal and an abnormal cell line, for example trisomy 21, tends to reduce the severity of the clinical presentation. Other mosaic karyotypes include a cell line with a mitotically unstable, structurally abnormal chromosome such as a ring, dicentric or ESAC.

Mosaicism is relatively rare except in specific circumstances. 45,X/46,XX mosaicism is found in up to 10% of females with Turner syndrome. Trisomy 8 is found in mosaic form in patients presenting with a syndrome involving specific dysmorphism. Pallister–Killian syndrome involves mosaicism for a supernumerary isochromosome composed of two copies of the short arm of chromosome 12, an abnormality never seen in lymphocyte preparations. Patients suspected of having this last syndrome are investigated by analysis of a skin biopsy.

A search for chromosome mosaicism may be made when a patient presents with asymmetrical growth, or hemihypertrophy: different proportions of normal and abnormal cells on one or other side of the body may lead to different growth rates.

Many ESACs identified to date originate from the region around the centromere of **chromosome 15**, and basically fall into two types: those which are small and benign, and those which are larger and occur in patients with a recognizable syndrome of mental handicap and physical dysmorphism. This latter type of chromosome is usually dicentric and contains the centromeres of both maternal chromosome 15s. It is believed that some kind of anomalous meiotic pairing and recombination gives rise to these abnormal chromosomes.

Increasing use of *in situ* hybridization may mean that mosaic abnormalities, especially those confined to particular tissues, can be confirmed more easily by using interphase nuclei recovered from buccal scrapes or hair roots. For example, use of a chromosome 8 centromere probe may show two populations of nuclei, one with two signals, one with three (see Chapter 6).

Sex chromosome imbalance is relatively well tolerated in comparison with autosomal imbalance. The most common karyotype in Turner syndrome has 45 chromosomes including a single X, with mosaic, ring and isochromosome variants. Turner syndrome occurs in approximately one in 2500 females, with a phenotype consistently including short stature, lack of secondary sexual development, and rudimentary 'streak' ovaries. Some patients have redundant skin around the neck, sometimes called neck webbing, a consequence of foetal oedema where there is a build-up of fluid in the lymphatic system,

Special properties of the sex chromosomes

Male sexual differentiation is controlled by the Y chromosome; in the absence of the Y, or indeed absence of the *SRY* male determining gene, female foetal development will occur. However, normal female development, postnatal growth and secondary sexual differentiation depend on the presence of at least two X chromosomes.

The Y chromosome includes few genes concerned with processes other than sexual differentiation and stature, while the larger X chromosome includes many genes unrelated directly to sexual differentiation or processes. Thus females are disomic for many X-linked genes for which males are monosomic and, in order to compensate for this difference, one X chromosome in the normal female becomes inactivated early in embryonic development. This process of inactivation is sometimes referred to as **lyonization**, after the cytogeneticist Mary F. Lyon who first proposed the mechanism. The inactive X chromosome replicates at the end of the S phase of the cell cycle and, providing both X chromosomes are normal, X-inactivation is random, with half of the cells expressing the maternal X-linked genes and half the paternal. The gene controlling the lyonization process, the X-inactivation centre, is located on the long arm of the X, and structurally abnormal chromosomes which incorporate X material can only be inactivated if the inactivation centre is present (Chapter 2). When the X chromosome is involved in a structural rearrangement, X-inactivation is frequently non-random, since those cells that are effectively more genetically balanced have a selective advantage.

X;autosome translocations

This type of translocation is very rare, especially in males. In females, the balanced carrier may be infertile if the X chromosome breakpoint occurs in a segment of the long arm known as the **critical region.** Sometimes a female who has a partial autosomal trisomy as a result of an unbalanced X;autosome translocation may display relatively mild clinical features, since the lyonization of the abnormal X spreads into the segment of the autosome and partially inactivates it.

Table 5.4 gives examples of X-inactivation in sex chromosome abnormalities.

XY pairing

Although the X and Y chromosomes are very different in morphology and genetic composition, they do have some genes in common and, at meiosis, there is recombination between homologous DNA sequences located at the tip of the short arm of each sex chromosome. The obligatory crossover between the X and Y is an

Table 5.4 *X-inactivation in sex chromosome abnormalities*

Phenotypic sex	*Chromosome complement*	*No. of inactive X chromosomes*	*Pattern*
Male	XY	None	
	XXY	One	Random
	XXXY	Two	Random
Female	XX	One	Random
	XXX	Two	Random
	X,del(X)	One	Abnormal X inactivated
	X,i(X)	One	Abnormal X inactivated
	Balanced t(X;autosome)	One	Normal X inactivated
	Unbalanced t(X;autosome)	One	Usually abnormal X inactivated

important factor in controlling the segregation of the two sex chromosomes into separate daughter cells at the reduction division, meiosis 1.

Sex reversal

There are examples of persons with a gender contradictory to their chromosomal sex. Females with a 46,XY karyotype include those with testicular feminization syndrome where the gonads are testes but a deficiency of cell binding of testosterone inhibits male differentiation. In pure gonadal dysgenesis there is no testicular differentiation and hence no production of testosterone, no male differentiation and no secondary sexual development.

Males with a 46,XX karyotype mostly have a cryptic translocation so that the critical male determining region of the Y chromosome is, in fact, present. Most of these males resemble Klinefelter's syndrome, and most of them can be shown by detailed molecular analysis to have undergone an anomalous meiotic pairing between the X and the Y, such that the male determining factors are translocated to the tip of the short arm of the X chromosome (see also Chapter 3).

Rare cases of males with a 45,X karyotype are usually attributable to the unbalanced translocation of the male determining part of the Y to the tip of an autosome.

particularly in the neck region, causing a fluid filled cyst. About 20% have a heart defect. Intelligence and lifespan are normal.

Approximately one in 1000 males has the karyotype 47,XXY, which gives rise to Klinefelter syndrome. XXY males do not produce adult levels of testosterone, and often have poorly developed secondary sexual characteristics and small testes. Some have gynaecomastia (breast development), and as a group they are tall with long limbs. The patient is often identified through the infertility clinic, although approximately 20% display mild mental handicap which may be noted during childhood.

An XYY karyotype occurs in about one male in 1000. Often asymptomatic with normal fertility, these males tend to be tall. In surveys of high-security criminal hospitals in the 1960s, more of these males were found than would be expected. At the time, these observations led to a popular misconception that the XYY karyotype was found in big men prone to aggressive and violent behaviour. This has turned out to be untrue, although when these boys are compared to their brothers, there does appear to be a tendency towards relatively low IQ, poor social adaptation, disruptive behaviour, and so on.

Suggested further reading

Gardner, R.J.M. and Sutherland, G.R. (1996). *Chromosome Abnormalities and Genetic Counselling*, 2nd Edn. Oxford University Press.

Replication banding, a valuable method for visualization of the late replicating X chromosome, exploits the properties of the substance 5'bromodeoxyuridine, or 5'BrdU, which has been mentioned before in relation to sister chromatid exchange (see Chapter 3). The G_2 phase is known to be about 4–5 hours, so addition of 5'BrdU to a culture for the final 6–7 hours before harvest ensures that it is incorporated only into the late replicating regions of the genome, which can then be differentiated using particular stains and buffers.

Rooney, D.E. and Czepulkowski, B.H. (1992). *Human Cytogenetics, A Practical Approach. Volume I, Constitutional Analysis.* Oxford University Press.

Self-assessment questions

1. Which chromosome abnormality would you expect to find in: (a) a Down syndrome child born to a 40 year old mother; (b) a Down syndrome child with a similarly affected brother and maternal aunt?
2. How many chromosomes are most commonly found in the diploid karyotype in Edwards, cri du chat, and Turner syndromes respectively?
3. If a normal chromosome has the structure ABCDEFG, and an inverted one is ABEDCFG, what are the consequences of recombination in the heterozygote?
4. What is: (a) a microdeletion; (b) the difference between terminal and interstitial deletions?
5. What sort of chromosome abnormalities contribute to the high level of spontaneous abortion?
6. Explain the difference between a reciprocal and a Robertsonian translocation.

Key Concepts and Facts

Definition

- Cytogenetics is the microscopic study of chromosomes.

Laboratory Techniques and Skills

- As chromosomes are only visible at mitosis, most analysis is undertaken on cells that have been grown in tissue culture, such as blood lymphocytes, amniotic fluid cells and skin cells. Chromosome analysis makes use of high-power oil-immersion optics.

Effects of Unbalanced and Balanced Karyotypes

- An unbalanced karyotype is generally responsible for serious clinical effects. Common unbalanced karyotypes include trisomy, deletion and products of segregation of a translocation. The carrier of a balanced structural chromosome abnormality frequently has a risk of miscarriage or birth of a physically and mentally handicapped child. Common balanced structural chromosome abnormalities conferring a reproductive risk include Robertsonian and reciprocal translocations, and pericentric inversions.

Impact on Sexual Differentiation and Development

- The Y chromosome contains the male determining gene essential for male differentiation. Sex chromosome abnormalities are associated with their own particular syndromes, and imbalance of the sex chromosomes is tolerated to a greater extent than autosomal imbalance.

Chapter 6
Molecular cytogenetics

Learning objectives

After studying this chapter you should confidently be able to:

State the limits of resolution of ISH.

Describe the role of FISH in the clinical cytogenetics laboratory, including advantages over routine cytogenetics.

Outline the main steps in a typical FISH procedure.

List some examples of FISH in clinical use.

Describe some different FISH systems.

Discuss other uses of FISH in genetics.

In the last chapter we have seen how the detection of chromosomal abnormalities depends on the limits of resolution of the light microscope, which we have already determined as between three and five Mb of DNA. By using the techniques found in molecular cytogenetics the resolution can be increased to between 10 kb and 1 Mb. This is achieved by combining two techniques: the cytogenetic process of producing metaphase or interphase preparations on slides, and that of recombinant DNA technology. Molecular cytogenetics may therefore be used when conventional cytogenetics has failed to detect a chromosome abnormality which may have been suspected due to the clinical reasons for referral. Another major advantage of the technique is that good chromosome preparations are not necessarily required; even interphase cells can be used in certain conditions.

The usual name for molecular cytogenetics is ***in situ* hybridization (ISH).**

Probes

It is possible to design a short piece of DNA complementary to a region of interest on a particular chromosome. This piece of DNA is called a **probe.** The DNA probes used in ISH are usually 2–55 kb in size.

The length of DNA which represents the human probe is often called an **insert**, because it cannot be copied in isolation, but must be inserted into another (often circular) piece of DNA called a **vector** (or cloning vehicle). This carries the human insert through all of its subsequent manipulations.

Vectors are in turn usually found in bacteria or yeast (the **host**), which can be easily grown up in the laboratory; as the bacterium divides, so does the vector and its insert, and thus many more copies are produced. This is referred to as **cloning.**

A particular example of a probe used in ISH is the 'cosmid contig'. A cosmid is a vector comprising a large circular piece of DNA which can carry a correspondingly large piece of human DNA (the probe).

The word 'contig' derives from the word 'contiguous' meaning 'next to'. Despite this, the probes usually map to overlapping areas of a DNA sequence.

Basic principles of ISH

Chromosome preparations are made in the routine way from an appropriate tissue (see Chapter 5).

The patient's chromosomal DNA on the slide is denatured to make it single stranded. The probe is labelled to make a signal visible and is then denatured so that it is also single stranded.

The probe solution is then added on to the surface of the slide. As the probe DNA has a very specific complementary sequence, it will find that matching sequence on the patient's DNA and anneal (**hybridize**) to it. The signal can then be detected.

Figure 6.1 shows not only the general principles, but also more technical details of a particular method of ISH known as **f**luorescent **i**n ***s**itu* **h**ybridization (**FISH**), which has become the method of choice in most laboratories.

Steps in a typical FISH procedure

(1) The slide is immersed in a solution of RNase, used to improve probe access.
(2) The slide is then taken through a series of alcohols (e.g. 70%, 80% and 95%) which dehydrate the slide and again will improve probe access.
(3) The DNA of the chromosomes is made single stranded by immersing the slide in a solution of formamide and salt solution at 70–75°C. This is called **denaturation.**
(4) The slide is then placed in a container of chilled 70% ethanol. This is to 'snap freeze' the single-stranded DNA and prevent it reannealing into its double-stranded form. More dehydration in 80%, 90% and 100% ethanol solutions follow.
(5) The probe, which will already be labelled with a base analogue of DNA, is heated at 70–75°C in order to make it

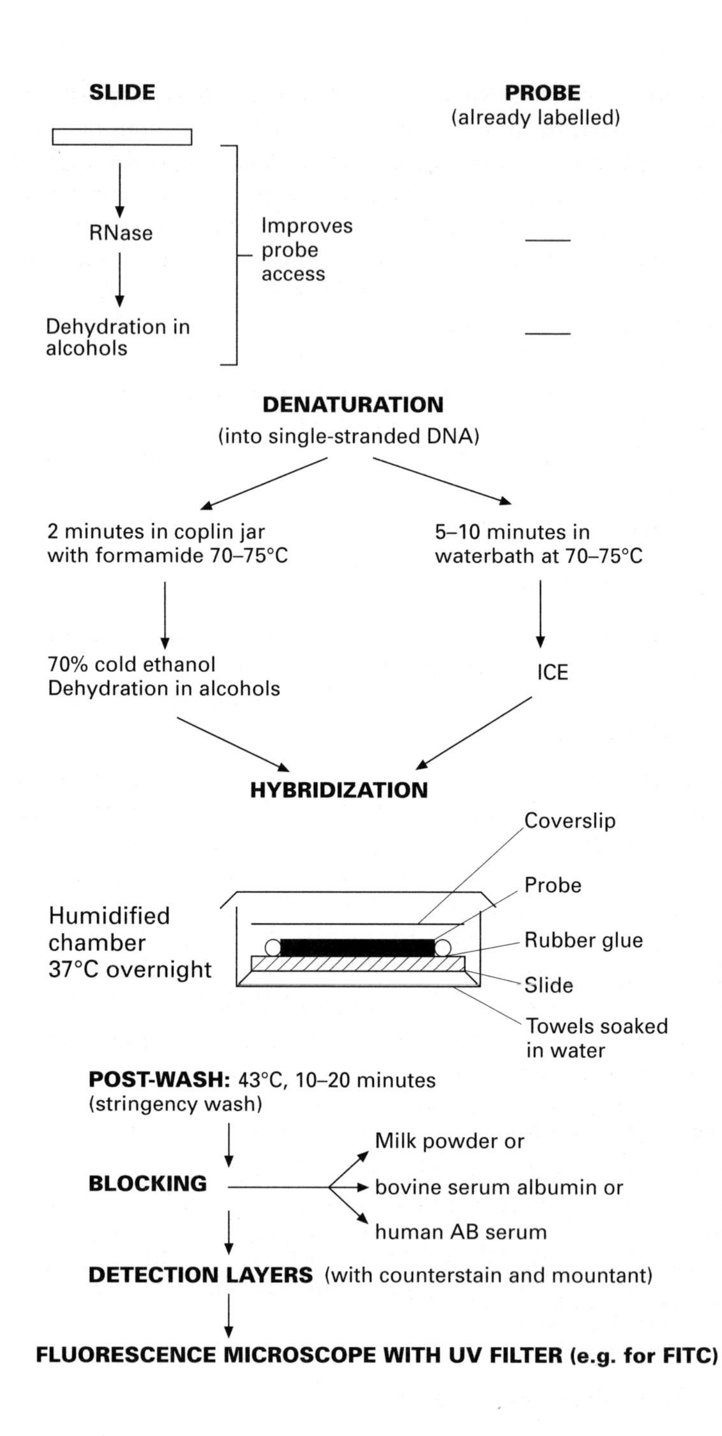

Figure 6.1 *General principles of* in situ *hybridization*

single stranded. To prevent reannealing, the tube is placed on ice.

(6) A drop of the single-stranded probe is placed on to the slide bearing the single-stranded chromosomal DNA and sealed beneath a coverslip, which is incubated overnight at 37°C in

a humidified chamber. This is the important **hybridization** step.

(7) It is most important to achieve a balance between washing away any unattached probe and leaving a clean preparation with only the well-matched probe hybridized to its complementary DNA. This stage is the **post-wash**, and often comprises a solution of formamide and salt at a particular concentration and temperature. This is the correct **stringency** of solution.

(8) The area of hybridization must be visualized using various **detection layers** which are dictated by the label on the probe. An example is given for the label **biotin** in Figure 6.2, which uses a **fluorescent** signal.

(9) A counterstain is now applied to make the chromosomes visible, after which a mountant and coverslip are placed on the slide.

(10) As a fluorescent signal is used, the slide must be examined under a fluorescence microscope, which comprises a light microscope with a particular filter set appropriate for the fluorescent dyes used.

The detection layers for the label biotin are also depicted in Figure 6.2; if we examine the theory more closely, this will provide a basis for understanding any system the student may encounter.

In order to make the signal from a probe visible, the final stage must involve the attachment of a dye to the probe. Why should the dye attach to the probe DNA, and how is the probe modified in order to achieve this?

The first stage is to **label** the probe with a reporter molecule. In this example we are using **biotin**, a naturally occurring cellular compound. The biotin has been modified with a linker arm such that it resembles one of the bases of DNA – in this case biotin-16-dUTP, a base analogue of thymine.

The labelling of the probe for use in FISH procedures is usually done by a process known as **nick translation** (Figure 6.3). Although there is another method of labelling which we will meet in the next chapter, nick translation is used as it is capable of labelling the large amounts of probe needed for FISH.

Stringency is increased (i.e. washes more probe off) as temperature is increased, but as the salt concentration is decreased. To decrease the stringency, one would either decrease the temperature or (more usually) increase the salt concentration.

The salt is in fact a mixture of two salts: sodium chloride and trisodium citrate (SSC).

Formamide is used in the hybridizing step and sometimes in the post-wash step as it alters the melting temperature of the DNA duplex. As the concentration of formamide in a solution is increased, the temperature needed to denature DNA is decreased. Instead of having to boil the slides or probe, they can be denatured at about 70°C.

As the probe has a tendency to try and stick to other substances on the slide (protein for example), there must be a **blocking** step which will 'mop up' the human protein, and ensure a cleaner preparation. Various substances may be combined with the detection layers, such as human AB serum, BSA (bovine serum albumin), or even milk powder.

A counterstain is a second fluorescent stain of contrasting colour.

Signal amplification procedure

Once the probe is hybridized to the complementary DNA on the chromosomes, we have to add the layers of fluorescent dye which will finally be detected as a signal (see Figure 6.2). This is done as follows:

(1) A layer of avidin is added to the slide, where it attaches to the biotin. A layer of FITC–avidin conjugate is added to the slide,

The disadvantage of forward painting is trying to decide which probe to try first, and perhaps going on to try several different probes, which is costly in terms of expense and time. To circumvent this problem, commercial **multiprobe systems** are available.

One type of kit contains a template slide divided into 24 areas upon which is pipetted cell suspension containing metaphase chromosomes. This is inverted over a device comprising corresponding chambers each containing pairs of 24 subtelomeric probes representing the individual chromosomes. Denaturation and hybridization follow. The probes for the short and long arms are labelled with different colours (one colour for each end).

Another limitation of chromosome painting is the difficulty in applying the technique to non-specific clinical referrals such as 'miscarriages', 'mental handicap' or dysmorphism, unlike conventional cytogenetics in which all the chromosomes are routinely analysed.

Denaturing a centromeric probe is simply achieved by heating to around 70°C for 5 minutes before applying it to the slide. Painting probes are also denatured at 70°C for 10 minutes, but are then placed in a waterbath for 1 hour at 37°C before being applied to the slide. This would not appear to make sense, as denatured DNA left unchilled usually reanneals back into double strands. The long reannealing time allows chromosome *in situ* suppression to occur.

Probes hybridizing to entire chromosomes (chromosome paints)

Sufficient unique sequence probes are cloned to give a representative coverage of a particular chromosome at many different loci. Once they are labelled and pooled, the effect will be to hybridize with and light up the whole of that chromosome. These are called **painting** probes, and the resulting FISH technique is called **chromosome painting.**

Chromosome painting

The routine technique described above is sometimes called **forward painting,** when a pooled **probe derived from normal human sequences** is hybridized on to a slide bearing metaphases with an **abnormal karyotype,** in an attempt to light up the areas of abnormality and deduce their derivation.

Although in principle a chromosome paint should cover the entire chromosome, in practice there sometimes appear to be unpainted areas or 'gaps' which remain 'uncoloured'. These are usually the regions containing certain repetitive or satellite DNAs such as the centromere, heterochromatic regions (e.g. 9q, 1q) or the telomeres. The unique sequence pooled probes are not designed to anneal with common repetitive sequences such as these.

Chromosome *in situ* suppression (CISS)

To improve visualization of painting probes, a technique known as **chromosome *in situ* suppression** is used (Figure 6.7).

Uses of chromosome paints

As with the previous categories of probes, paints are used to **identify chromosomes** and to **detect chromosomal abnormalities.** Whereas the subregional probes tend to be of limited use, mostly in numerical analysis, the painting probes have a wider range of use, especially with respect to **structural aberrations**. They can be used appropriately in the following circumstances:

- **Identification of the origin of additional material** in an unbalanced translocation or a duplication.
- **Rapid identification of translocated chromosomes in prenatal diagnosis** where the level of resolution of the chromosomes in prenatal material such as chorion villus biopsies is not satisfactory.
- To **improve the interpretation** of the complex chromosome changes observed in metaphases from **tumours.**

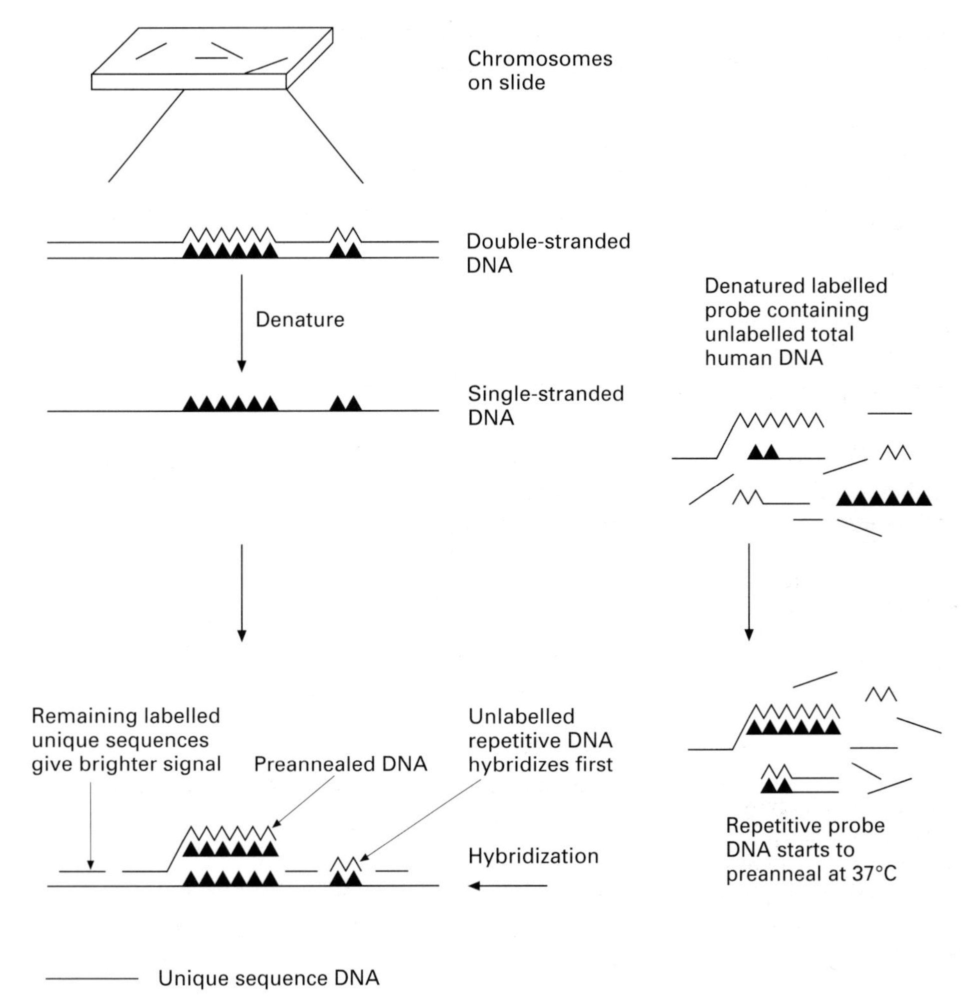

Figure 6.7 *Chromosome* in situ *suppression*

- **Pseudoautosomal studies** using a Y paint on an X chromosome or vice versa.

Reverse painting

In reverse painting, in contrast to the previously described technique (forward painting), a **probe is made from the abnormal chromosome.** It may be from a *de novo* translocation, a complex unbalanced rearrangement, or more usually a marker of uncertain origin. The probe is hybridized on to metaphases with a **normal karyotype.** The derivation of the abnormal chromosome(s)

Painting probes contain **competitor DNA**, which is total human DNA, either derived from human placental DNA or a particular human satellite DNA called cot-1 DNA.

The chromosomal DNA on the slide and sometimes the probe DNA itself contain repetitive sequences. We have already seen how gaps may arise due to this feature, but the **whole** of a painted **chromosome** may appear very faint due to intrachromosomal repetitive sequences.

If competitor DNA is introduced via the probe, the repetitive sequences in the competitor DNA will hybridize with the repetitive sequences in the probe very quickly, leaving unique sequence probe ready to anneal to the chromosomal DNA. When the probe is added to the slide, the repeats in the competitor DNA also reanneal quickly to the repeats in the chromosomal DNA. This leaves more target (unique) sequence chromosomal DNA ready to receive the unique sequences of the preblocked probe. The result is a brighter signal with less background cross-hybridization than could be obtained without preblocking.

The abnormal probes used in reverse painting can be derived in two ways:

(1) **Microdissection**: Microdissection apparatus comprises mechanical controls attached to the stage of a microscope that enable an operator to use a finely drawn glass pipette with a chiselled end to slice the suspected abnormal chromosome or marker and lift it from the slide. Using as few as thirty pieces of chromosome, a probe can be made from the chromosomal DNA by the molecular technique known as the polymerase chain reaction (PCR).

(2) **Flow sorting**: This method is also called flow cytometry. A suspension of chromosomes is made from the abnormal karyotype. A machine called a **fluores-**

involved can then easily be seen, as parts of the relevant normal chromosomes will light up.

Other FISH techniques

Comparative genomic hybridization (CGH)

This technique is usually applied to situations where a karyotype may be particularly difficult to elucidate due to multiple or unusual abnormalities.

The basis of the technique is to compare the test DNA with normal DNA. For example, if a tumour karyotype included trisomy 1, then there would be one and a half times as much chromosome 1 DNA arising from the tumour as there would be in a normal karyotype (see Figure 6.8 for details).

The disadvantage of the technique is that it cannot detect balanced rearrangements such as translocations, as the total amounts of DNA would be the same.

The advantage is that if more than one unbalanced abnormality is present in the same metaphase (which is often the case in tumours), then CGH can display all the imbalances at once, in one test.

Molecular FISH techniques

The polymerase chain reaction (PCR) has been an essential part of molecular biology since 1985. It enables a very small amount of DNA to be copied using a particular enzyme which works optimally at the high temperatures used in the reaction.

Although this technique is discussed more thoroughly in Chapter 7, elements of PCR are exploited in some of the more sophisticated FISH techniques explained below.

Primed *in situ* hybridization (PRINS) and *in situ* PCR

If a short piece of single-stranded DNA (a primer) is designed complementary to a centromeric repeat or a known unique sequence of a single gene locus, these areas can be copied using a polymerase enzyme. There is always a denaturing step at around 94°C to denature the chromosomal DNA, a primer annealing step at about 55–65°C, and an extension step at around 72°C. This is equivalent to one cycle of a PCR reaction and is known as **primed in *situ*** hybridization (**PRINS**).

The advantage over a 'probe-based' FISH technique is that there is no excess probe which may cause cross-hybridization or give undesirable background fluorescence. The signal is absolutely specific to the area where the incorporation of the reporter molecule took place, which should give a cleaner signal with less background.

In practice, however, the signal is often rather faint. By modifying PRINS such that specific temperatures are repeated for 28 to 35 cycles, a 'layer' of target DNA can be built up at the site of interest.

This will obviously produce a **brighter signal** as detection layers are added, and is known as ***in situ*** **PCR**.

If a base analogue directly conjugated with a fluorescent dye (such as FITC) is incorporated during the PCR reaction, then visualization will be accomplished in a single step during the PCR. This **improves the detection time** of the technique, sometimes by several hours.

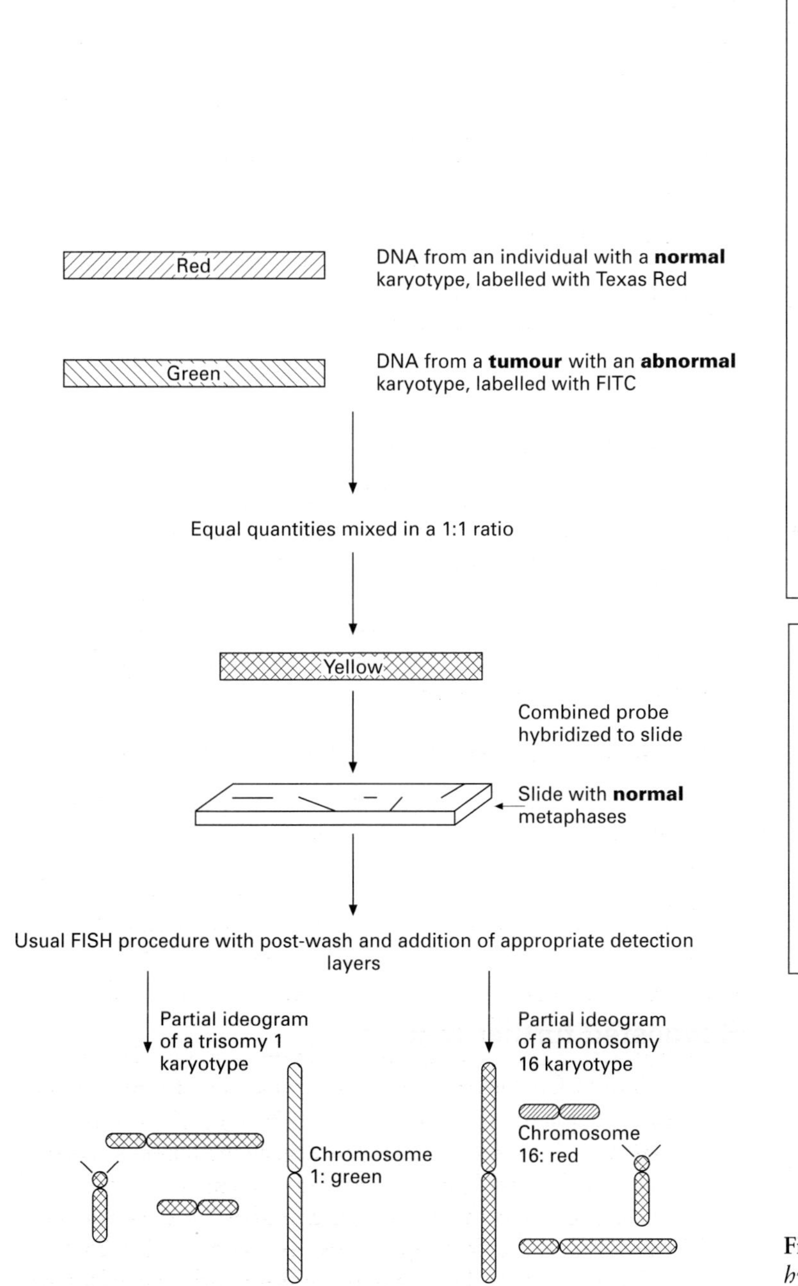

cent activated cell sorter (FACS) is used to detect the aberrant marker chromosome and produce an enriched supply which can be converted into DNA and used as a probe. The standard method for making normal chromosome paints is also using a FACS. The sorter comprises a fine laser beam, which shines through a stream of fluorescently stained chromosomes derived from the abnormal karyotype. These are separated by size, and computer software is used to divert and isolate specific markers or other chromosomes of interest. There are problems with any abnormal chromosomes that are similar in size to normal chromosomes, and the final result would not be a pure solution of abnormal marker DNA, as it will be contaminated with other chromosomes of similar size.

Solid tumours frequently tolerate both structural and numerical abnormalities, and are therefore difficult to analyse by conventional means. It is very difficult to grow cultures or obtain chromosome preparations from many types of tumour, so **CGH is an effective strategy as it uses genomic DNA rather than metaphase preparations**.

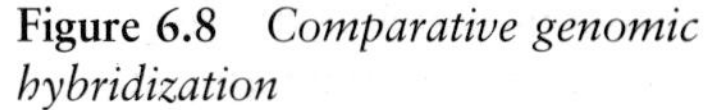

Figure 6.8 *Comparative genomic hybridization*

A PCR solution of the specific single-stranded **primer** or sometimes a double-stranded oligonucleotide (usually 18–30 bp), together with **dATP, dCTP, dGTP and dTTP** (one of which is directly conjugated with a **fluorescent dye**) and the heat-resistant enzyme **taq polymerase**, are added to the chromosomes on the slide. The chromosomal DNA is denatured (if formamide is present this will occur at 70°C) and then, as the temperature is lowered, the primer anneals to its complementary sequence to form a double-stranded template. The taq polymerase 'logs on' to the template and starts to copy the appropriate sequence (the temperature is raised to 72°C to allow maximum extension by the enzyme).

By repeating these cycles of temperatures the two strands of DNA are copied exponentially until hundreds of thousands of similar sized pieces of DNA are produced.

It would be convenient to amplify unknown sequences of DNA (such as markers) which could then be used as probes in reverse painting. As the whole premise of PCR is based on primers being designed from known sequences of DNA, it would seem unlikely that markers derived by microdissection or flow sorting would be able to be amplified, as their origin is not known.

Degenerate oligonucleotide primed-PCR (DOP-PCR) uses two modifications of routine PCR to circumvent this problem. The single primer (CCG ACT CGA GNN NNN NAT GTG G) has a central section comprising six random nucleotides (N). This ensures that each random

Other applications of FISH

Gene mapping

FISH is a very efficient way of finding the locus of a new probe, even if only the band location is determined initially.

Once FISH gives an approximate probe location, other chromosome walking or jumping techniques can refine the specific site.

With the application of new chromosome spreading techniques, we can now look at long pro-metaphase chromosomes, and even at chromosomes in interphase, which have very little condensation, and are therefore extremely elongated if drawn out on the slide.

If the locus order of several novel probes needs to be elucidated, each probe can be labelled with a different reporter molecule, and hence a different colour. The resultant **fibre FISH** shows a long strand of chromosomal DNA with the coloured probes attached in order, like beads on a string.

Even the production of colour no longer depends on independent reporter molecules. In CGH (Figure 6.8) we saw that an equal mixture of red and green fluorescent dyes could give a third colour – yellow. If different proportions of red and green are used (e.g. 1 : 4, 1 : 1, 4 : 1), then (together with the pure red and green), a computer can be programmed to see five different artificially assigned colours; the imaging software now provides colour detection beyond the resolution of the human eye.

Summary

The development of FISH techniques has permitted the identification of small deletions not previously visible and that of unknown markers and rearrangements. The techniques now available even provide information from uncultured (interphase) cells at a resolution not previously possible.

There is no doubt that the number of probes and the number of colours will continue to increase, enabling the laboratory to provide more information to the clinicians, and ultimately improving both prenatal and postnatal diagnosis and hence patient care.

Suggested further reading

Lichter, P. and Cremer, T. (1992). Chromosome analysis by non-isotopic *in situ* hybridization, in *Human Cytogenetics: A Practical Approach, Volume 1* (eds D.E. Rooney and B.H. Czepulkowski). IRL Press at Oxford University Press.

Ward, D.C., Boyle, A. and Haaf, T. (1995). Fluorescence *in situ* hybridization techniques, in *Human Chromosomes: Principles and Techniques* (eds R.S. Verma and A. Babu). McGraw-Hill.

Self-assessment questions

1. What is a probe?
2. Name three uses of FISH on metaphase cells.
3. Name three clinical syndromes detected by FISH and two features of each (use the Appendix).
4. After performing FISH, your slide has a lot of background. Define stringency and explain what the result would be of increasing the stringency.
5. What are the limits of resolution of FISH?
6. A little girl presents with absent speech, very stiff limbs and severe mental retardation. Her chromosomes appear normal by conventional microscopy. Which FISH probe might you try, and what might you expect to see?
7. What is the difference between a dual-pass and a triple-pass filter?
8. An ultrasound scan reveals a foetus to have a heart defect. Amniotic fluid cells are available. Which FISH probe might you use and why?

sequence has a chance of finding a complementary sequence somewhere in the denatured marker DNA.

Rather than using a primer annealing temperature of 55–66°C, the DOP-PCR programme lowers the annealing temperature to 30°C, which ensures that any loosely hybridized primer will remain annealed and copied. After a PCR of five cycles, some of the product is taken in order to PCR more conventionally.

This application is also useful for making copies of any DNA probes where the sequence is not fully known.

FISH has proved very useful in the study of fragile sites. If a series of probes from the area of the fragile site are available, they can be visually mapped to either side of a specific fragile site by FISH. If another fragile site is suspected to be very close to the first one, the pattern of the probes can confirm the same order as before, or reveal a new independent site as the probe markers will be on different sides of the site.

This was illustrated with the Xq27.3 fragile X site called FRAXA, associated with the fragile X syndrome. When another site at a very similar position was discovered, the ordering of probes on either side of the new site was different to FRAXA. This was the new fragile X syndrome FRAXE, the locus of which was 600 Mb distal to FRAXA.

M FISH or **multiplex** FISH uses four to five fluorochromes and the application of computer software to generate 24 colours, such that a separate colour can be assigned to each chromosome.

R_x **FISH** or **cross-species FISH** uses paints from gibbon chromosomes labelled with different coloured fluorochromes, hybridized to human chromosomes. This results in a multi-coloured pattern of about 100 bands.

Although human probes should be unique to human DNA, many genes have been so important in evolution that the sequences have remained unchanged for millions of years. In this case human probes would be able to detect analogous **conserved** sequences in other species such as the primates. The number of conserved sequences we share with other animals also gives us an idea of how long ago we diverged from a particular line, as the number of accumulated mutations will increase with time.

Even just looking at the karyotype of a chimpanzee makes it easier to appreciate that humans share 98.5% homology with this species. Many of their chromosomes are recognizable, and many others are derived from inversions or translocations.

If preparations of chromosomes are made from a non-human species (e.g. chimpanzee or orang-utan), and human probe DNA is applied, then the extent of cross-hybridization is readily seen.

Key Concepts and Facts

Advantages Over Conventional Cytogenetics

- Using FISH, resolution can be increased to kilobases rather than megabases.

Principles of FISH

- The basic steps include the denaturation of both slide and labelled probe, hybridization of the probe to the chromosomes on the slide, washing off of the excess unbound probe, and detection (with amplification) of the signal.

Modifications

- Two or more different reporter molecules enable more than one colour to be used.

Limitations

- Fluorescent probes are not as sensitive as radioactive probes, and may require amplification. Fluorescent dyes may fade quickly.

Applications

- FISH can be used on cultured or uncultured cells at interphase.
- Different kinds of probes are available, such that unique sequence probes may detect microdeletions, repetitive probes can be used for numerical abnormalities, and painting probes can identify structural aberrations.

Other FISH Techniques

- Improvements in FISH techniques enable more abnormalities to be detected simultaneously (CGH) or allow much faster visualization of the signal (PRINS and *in situ* PCR).

Chapter 7
Molecular genetics

Learning objectives

After studying this chapter you should confidently be able to:

Define the term probe and give examples of nomenclature.

Explain how restriction enzymes can reveal DNA polymorphisms.

Outline the principles of gel electrophoresis.

Describe the principles of Southern blotting and give examples of its use.

Explain gene tracking and informativeness.

Describe the principles of PCR and give examples of its use.

Outline some methods of mutation screening.

Genetics has always been regarded as a comparatively modern science; DNA structure was elucidated in 1953, and cytogenetics grew from the discovery of the correct number of human chromosomes by Tjio and Levan in 1956.

In the 1970s a whole new technology called molecular biology arose, which enabled the analysis of human DNA down to the single base pair level. By the 1980s it led to a new approach to screening for previously uncharacterized clinical disorders, and by the 1990s the techniques became automated such that the DNA sequence of the 46 human chromosomes is currently being elucidated, an endeavour entitled the Human Genome Project.

Molecular biology has now been incorporated into the study of inherited disorders and is known as **molecular genetics.**

Molecular genetic analysis is carried out for two reasons:

- Molecular genetics techniques may provide a more specific or more accurate result. Some molecular genetic tests provide a less expensive alternative to cytogenetic analysis, e.g. for fragile X syndrome and Prader–Willi/Angelman syndrome.
- As molecular genetics examines the DNA at the base pair level, certain syndromes which cannot be detected cytogenetically can be analysed for mutations or deletions (e.g. cystic fibrosis and

Duchenne muscular dystrophy) using molecular techniques, which has led to an understanding of the role of mutations in the generation of disease.

The two techniques on which most molecular analysis is based are **Southern blotting** and the **polymerase chain reaction (PCR)**.

Southern blotting

This is the older of the two techniques which, although superseded by PCR in the late 1980s, still plays a valuable role in molecular analysis.

The following four concepts provide the key to understanding Southern blotting:

- **Probes**
- **Restriction enzymes**
- **Gel electrophoresis**
- **Polymorphisms.**

DNA probes

Already briefly described in Chapter 6, these double-stranded pieces of DNA map to the gene or region of interest (usually the site of a genetic disorder). The probes used in Southern blotting are generally 0.3–5 kb.

A clinical laboratory rarely receives a probe as an independent solution of DNA fragments. The probe (or **insert**) is usually inserted into a double-stranded circular piece of DNA called a **plasmid** by enzyme cutting and pasting. As it carries the probe the plasmid is termed a **vector**.

The vector (which generally carries a gene for antibiotic resistance) and insert are exposed to bacteria which are **competent** to take up the plasmid. The bacteria are grown in broth containing an antibiotic to which the plasmid is resistant. The plasmids give the bacterium a selective advantage, and enable the plasmids and their inserts to divide also; this process enables large quantities of probe to be generated, and is called **cloning**.

Nomenclature

Probe names

Names may include the laboratory of origin or the probe size. Ox1.9 is a fragile X probe; it originated in Oxford and has a length of 1.9 kb. The name may indicate the number of attempts at production and the initials of the inventor or may have a prefix, e.g. 'p' for plasmid.

Locus name

The area on the human chromosome to which the probe is complementary is called the **locus**. The locus may therefore have a separate name which (unfortunately for the student) is often different from the probe name. Locus names often take the form given in the following example. In the locus 'name' D7S8: D = DNA, 7 = chromosome 7, S = unique segment and 8 = unique segment **number**. The cystic fibrosis probe pJ311 is complementary to the locus D7S8.

As several different probes may anneal to the same locus (which may cover a large area in molecular terms), molecular geneticists generally use the probe name. Older probes, or those hybridizing to

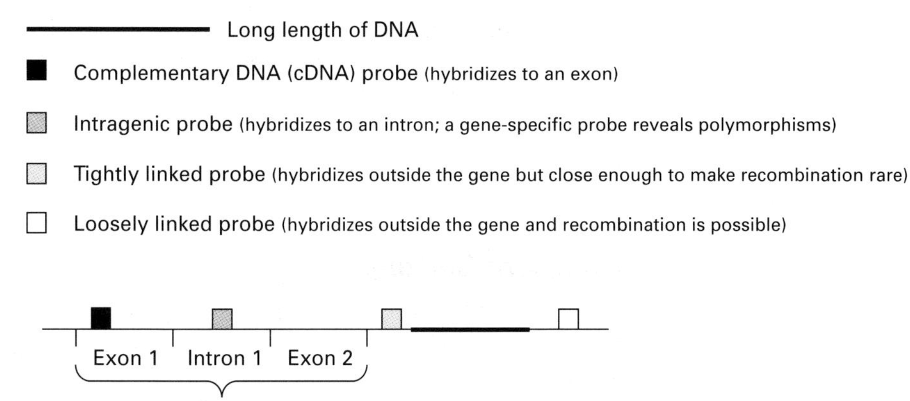

Figure 7.1 *Probe nomenclature and positions with respect to a gene*

Table 7.1 *Examples of probe nomenclature*

Locus name	*Probe name*	*Genetic disorder where used*
D22S75	N25	DiGeorge syndrome
KB17	SNRPN	Prader–Willi and Angelman syndromes
MET	MET D	Cystic fibrosis
D15S113	LS6-1	Prader–Willi and Angelman syndromes
D15S113	LS6-2	Prader–Willi and Angelman syndromes

an important critical region, may have the same locus name as probe name (see Table 7.1). Probes may also be described generically from their position with regard to the gene of interest (Figure 7.1).

Restriction enzymes and polymorphisms

In the early 1970s it was discovered that certain bacteria produce enzymes which cleave foreign DNA. These are now known as **restriction endonucleases**; the enzyme names reflect the bacterium of origin, for example:

- Taq I: *Thermus aquaticus*
- EcoRI: *E. coli* type ***R1***
- Hind III: *Haemophilus influenzae* type **d** III
- Msp I: *Moxarella* ***sp***.

Restriction enzymes (REs – as in Figure 7.2) cut double-stranded DNA at very specific base sequences (see definitions) called **restriction sites**, which occur many times in the genome. The resultant **digest** contains a mixture of differently sized **fragments** which can be separated by gel electrophoresis.

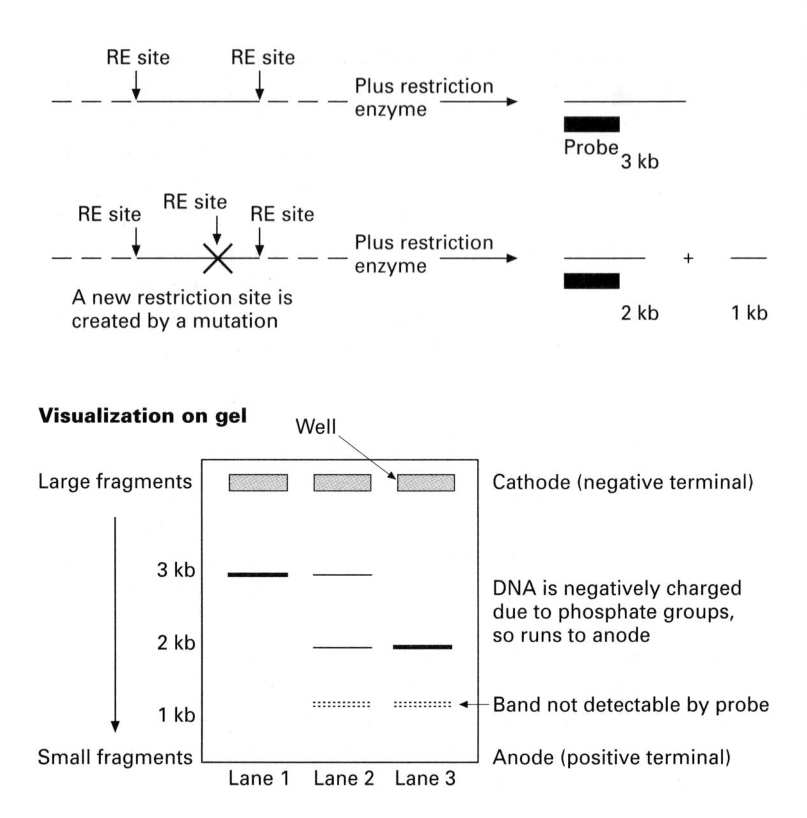

Figure 7.2 *Exploitation of RFLPs by gel electrophoresis*

Gels comprise a matrix of differently sized pores. They are either made from **agarose** powder which is dissolved by boiling in buffer and then sets 'mechanically' like jelly, or an **acrylamide** monomer is used which polymerizes by chemical means, using a catalyst. This is known as a **polyacrylamide** gel.

Gels are made and run in **buffer** which contains **ions** for **electrical conduction**.

They are either set in plastic trays (to run horizontally) or between glass plates (to run vertically). The DNA is loaded into **wells** which are created at one end of the gel using plastic combs with teeth of suitable sizes. The comb is removed and the tray and running buffer are placed in a gel tank.

Different percentage gels can be made depending on the size of the DNA fragments to be separated. A 0.8% agarose gel will separate fragments around 1–30 kb, a 2% agarose gel fragments around 100–500 bp, and a 13% polyacrylamide will separate gel fragments as little as 3 bp apart.

Gel electrophoresis

During gel electrophoresis the patient's restricted DNA fragments are subjected to an electrical current. As **DNA is negatively charged** due to the phosphate groups, the **DNA will migrate to the positive anode** of the gel tank (Figure 7.2).

Small DNA fragments run faster and therefore further than the larger fragments in a set time, and appear as bands nearer the lower end (anode) of the gel.

Polymorphisms

About every 100–200 bp in the introns of human DNA a natural variation in DNA sequence occurs. These mutations may be within the introns of genes, or within the intervening sequences between genes. If mutations occur in the conserved exons, the consequences could be serious. Over time, however, introns tend to accumulate mutations which will vary between different people, or even between the two chromosome homologues in one individual. These sequence differences are called **polymorphisms,** which can be detected using a probe complementary to the restriction site(s) of interest.

Restriction fragments of different sizes derived from introns are called **length polymorphisms**. The correct term for these digested fragments is therefore **restriction fragment length polymorphisms** or **RFLPs** for short. RFLPs are used in particular circumstances, often when the exact location of a gene is not known. Linked probes are then used, and the polymorphisms can be associated with a gene disorder and tracked through a family (see gene tracking).

Probes, which are some distance away from the gene, have disadvantages, as there is a chance of **recombination** between the locus at which the probe hybridizes and the gene of interest.

Principles of Southern blotting

The patient's double-stranded DNA is denatured by chemical treatment to make it single stranded. The double-stranded probe DNA is heat denatured and labelled. The label may be non-radioactive (using for example biotin, as in FISH) or radioactive (the isotope ^{32}P is often used in Southern blotting).

When the denatured probe is added to the single-stranded human DNA, it will find its complementary sequence and hybridize, producing a length of labelled double-stranded DNA. If it is non-radioactive the result is detected visually using dye (as in FISH). Radioactive methods use autoradiography. This relies on the radioactivity fogging a piece of X-ray film to produce black bands (Figure 7.3).

Steps involved in Southern blotting

The steps involve in Southern blotting are illustrated in Figure 7.4 and are described below.

DNA extraction

DNA can theoretically be extracted from any **nucleated cells**. In medical genetics the usual source is a blood sample containing nucleated white cells. Other sources include chorionic villi (derived from the placenta and hence the tissue is foetal in origin), amniotic fluid cells, tumours and histopathological paraffin wax preparations.

The nucleated cells are often treated with detergent, which breaks open the white cell membranes, and enzymes (proteases), which digest the proteins, thus leaving the DNA in an aqueous solution. This 'aqueous' solution is purified using an organic solution such as phenol. Chloroform is then used to remove the phenol from the DNA.

The DNA remaining in solution is precipitated using two volumes of ethanol (the ratio is quite strict – 2.5 volumes might precipitate RNA). The DNA appears as a white viscous substance

The pair of homologous chromosomes shown in Figure 7.2 display a sequence difference in an intron. On the first chromosome a 3 kb length of DNA is produced between two restriction sites. On the second chromosome a mutation has occurred such that an extra restriction site has been created. The enzyme now produces two differently sized pieces of DNA – one of 1 kb, and one of 2 kb. These will run to different levels on the gel.

The probe detects a 3 kb fragment from one chromosome and a 2 kb fragment on the other homologue (the probe has not hybridized to the 1 kb fragment, so although the band is present on the gel it will not be visualized).

Below are some definitions relevant to molecular genetics.

- **Restriction enzymes (endonucleases):** Naturally occurring endonucleases from bacteria, which cleave double-stranded DNA at specific base sequences, usually recognizing a four to six base sequence.

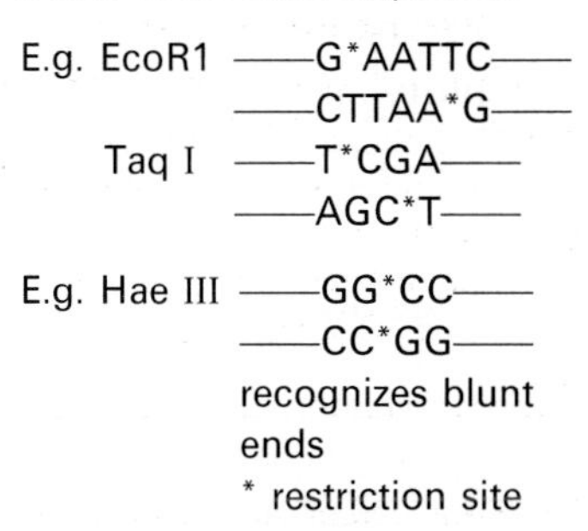

E.g. EcoR1 ——G*AATTC——
——CTTAA*G——
Taq I ——T*CGA——
——AGC*T——

E.g. Hae III ——GG*CC——
——CC*GG——
recognizes blunt ends
* restriction site

- **Restriction fragments:** The different lengths of DNA produced by a restriction enzyme.
- **Polymorphism:** In molecular genetics, this refers to a naturally occurring variation in a DNA sequence, usually in an intron.

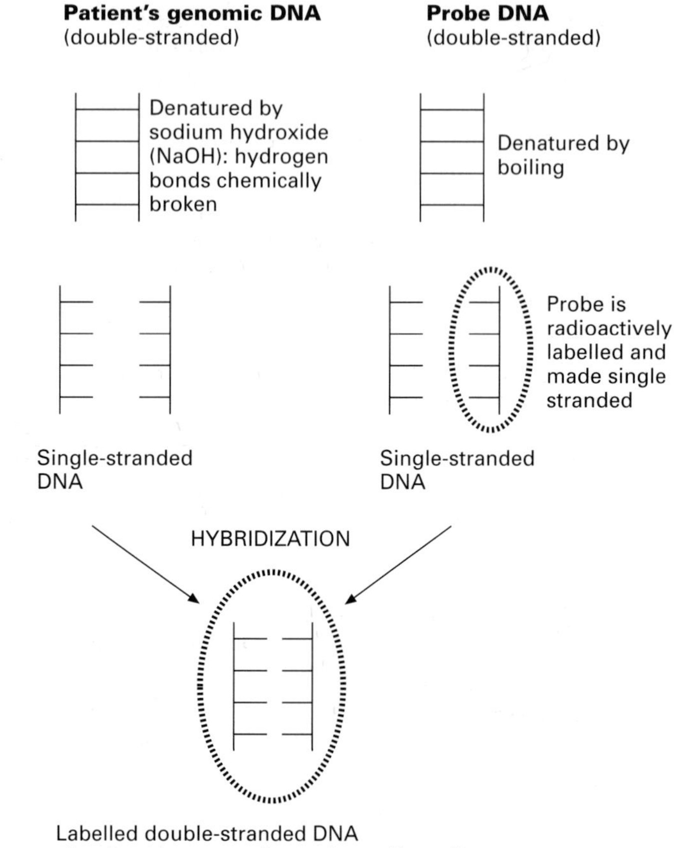

Figure 7.3 *Principles of Southern blotting*

- **RFLP (restriction fragment length polymorphism):** A length variation (i.e. polymorphism) in a DNA fragment, revealed by a restriction enzyme, due to the presence or absence of a restriction site.
- **Allele (general genetics):** The alternative form(s) of a gene at the same locus on homologous chromosomes.
- **Allele (molecular genetics):** The alternative length(s) of DNA (i.e. RFLPs) generated due to the presence or absence of a restriction site at the same locus on homologous chromosomes. Alleles may also be defined as alternative lengths of DNA (at the same locus on homologous chromosomes) detected by a probe or a pair of PCR primers.
- **DNA probe:** A short length of DNA, usually radiolabelled and made single stranded, which hybridizes with complementary DNA sequences linked to or containing the gene of interest.
- **Haplotype:** A combination of linked alleles inherited together as a unit (from one parent on one chromosome).
- **Linkage disequilibrium:** The association of two linked alleles more frequently than would be expected by chance.

Southern blotting detects specific sequences of DNA amongst the entire human genome using unique DNA probes. It is so sensitive that it can detect less than 0.1 pg of DNA – that is a single copy gene.

which can be hooked out of the ethanol, dried and dissolved in buffer or water.

DNA restriction digests

Following digestion of the total genomic DNA, the gene of interest will usually be present on a specific DNA fragment of known size.

Gel electrophoresis

The DNA fragments produced by a routine clinical digest vary between a few hundred base pairs and 50 kb in size, depending on the restriction site of the enzyme. The fragments are separated by gel electrophoresis; as each patient's DNA migrates from the well it leaves a streak in that particular track or **lane**.

The lanes of DNA are visualized under ultraviolet light by staining the gel with the fluorescent dye **ethidium bromide**, which intercalates into the major grooves of the DNA helix.

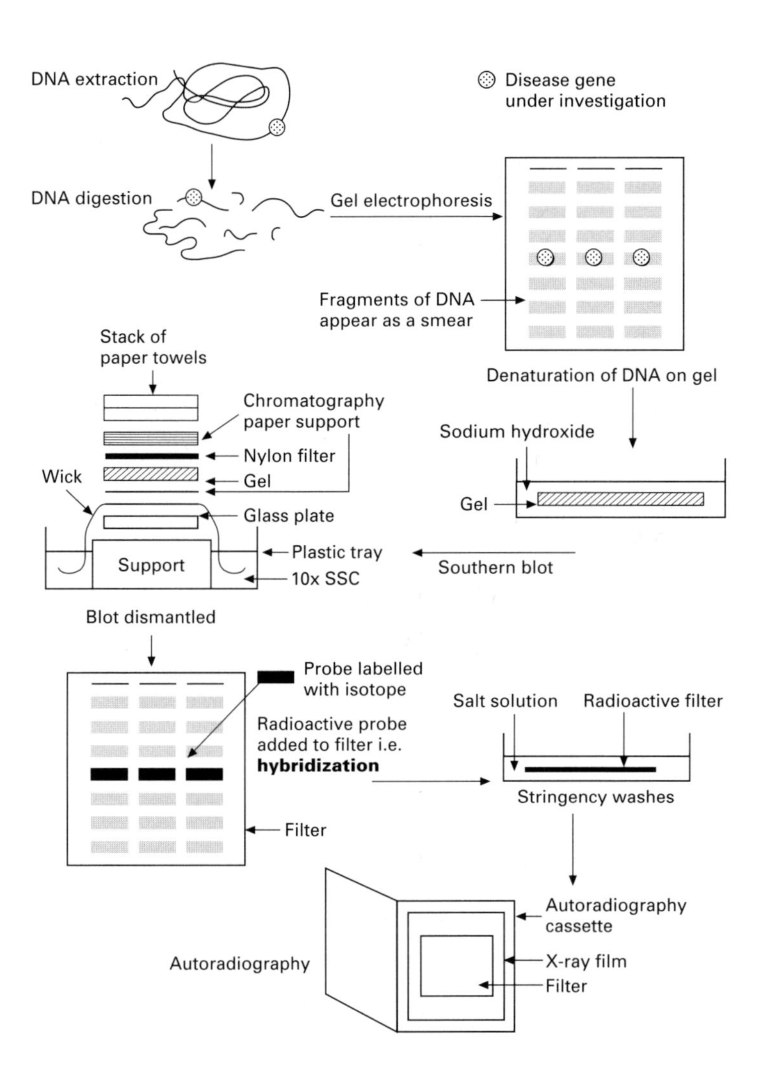

Figure 7.4 *Steps involved in Southern blotting*

Very small weights and volumes are used in DNA technology. The units are as follows:

Weights:

grams (g)	
milligrams	(1 mg = 10^{-3} g)
micrograms	(1 µg = 10^{-6} g)
nanograms	(1 ng = 10^{-9} g)
picograms	(1 pg = 10^{-12} g)

Volumes:

litre	(l)
millilitre	(1 ml = 10^{-3} l)
microlitre	(1 µl = 10^{-6} l)

Usually, 10 ml of whole blood yields about 500 µg of DNA, and a 3–4 mg chorionic villus sample (CVS) yields about 50 µg of DNA.

If contaminants such as red blood cells are present in a DNA extraction, the red cell membranes are **lysed** (broken open) leaving the white cells to be spun down. Wax preparations are treated with xylene until only tissue is left.

The concentration of the extracted DNA can be measured using a spectrophotometer. Nucleic acids absorb ultraviolet light at a specific wavelength of 260 nm, and this absorption can then be converted into a quantity of DNA. The concentration of DNA is usually expressed in µg/µl.

Denaturation

The DNA on the gel must be denatured in order to hybridize with the specific probe, which is added later.

Southern blot

As the denatured gel is too fragile for further manipulations, the DNA must be transferred on to a more durable membrane (sometimes called a filter) by the technique of Southern blotting. The salt solution is drawn by capillary action via the bridge (which acts like a wick), through the gel and into the dry paper towels. As it passes through the gel the DNA is transferred on to a nylon filter.

The DNA is permanently fixed or **cross-linked** to the membrane using ultraviolet light.

The restriction enzyme is not added all at once: DNA is so viscous that aliquots of enzyme are added – once an hour for 3 hours for example – so that it gradually penetrates the DNA.
Sometimes the enzyme does not cut at every available restriction enzyme site. This creates larger DNA fragments due to **partial digestion**.

Prehybridization

Prehybridization solution is added to the filter for a few hours or overnight, at a temperature determined by the amount of formamide in solution.

Labelling the probe

Radiolabelled probes are still commonly used as they are very sensitive. The usual method of labelling is **random primed labelling** (Figure 7.5), which has several advantages over the nick translation method used in FISH. The DNA insert (i.e. the probe) is cut from the vector before labelling.

Hybridization

The prehybridization solution is replaced with labelled probe, which has been reboiled and added to a small amount of hybridization solution.

Sometimes an initial step is performed where the gel is soaked in hydrochloric acid. This **depurination** step nicks the large molecular weight DNA, which will help it to transfer more easily at the Southern blotting stage.
The gel is then immersed in a tray of sodium hydroxide and gently shaken; this will break the hydrogen bonds and **denature** the patient's DNA.
A final **neutralization** step may be performed using 'TRIS' buffer; this will bring the pH back to neutral.

Stringency washes

These follow the same principles as for FISH; the stringency is empirically determined for each probe, and usually involves a specified **concentration of salt** solution (often SSC) at a particular temperature. The detergent **SDS** (sodium docecyl sulphate) is usually added to wash the isotope off more thoroughly.

Southern blots are named after their inventor Professor E. Southern working at Edinburgh. One of the most useful techniques in DNA was achieved using plastic trays and paper towels!

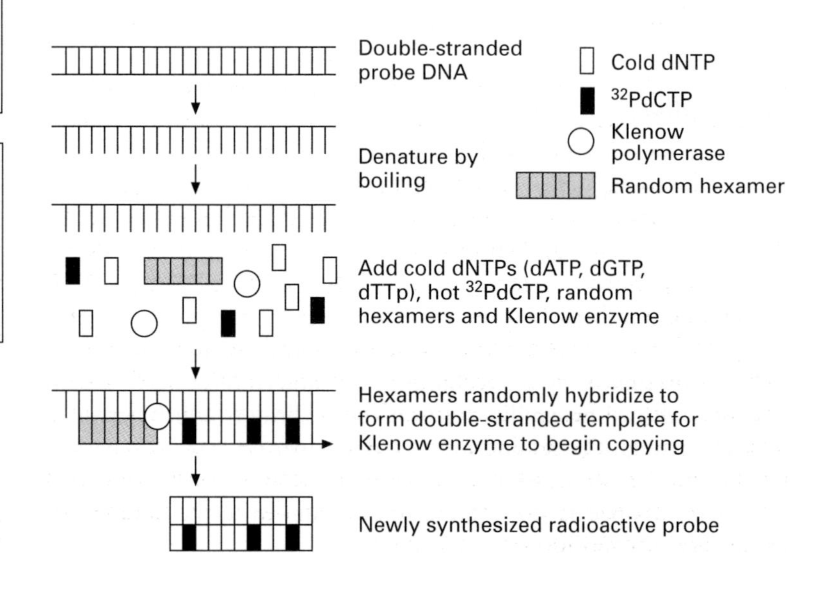

Figure 7.5 *Random primed labelling*

Autoradiography

In a suitable darkroom, the washed filter is wrapped in cling film and placed face down on a piece of medical X-ray film. The film is placed in a light-tight autoradiography cassette, and exposed at −70°C for 1–10 days. It is then developed; the X-ray film should display black bands where the hybridized radioactive probe has sent out hard beta particles to fog the X-ray film.

Most modern types of membrane are made of nylon, which is sometimes positively charged, so there is no need to cross-link the DNA. The older nitrocellulose membranes were very prone to grease marks and, after baking for two hours at 80°C, very brittle! As it is also very difficult to remove the first probe in order to rehybridize with a second probe, nitrocellulose has not been used in diagnostic clinical laboratories for many years.

Advantages of Southern blotting

- The exact gene location of a disease need not be known (see the sections on gene tracking and linked probes).
- It will detect large pieces of DNA.

Disadvantages of Southern blotting

- The process takes 7–14 days.
- It needs micrograms of DNA.
- Radioactivity has exposure limits.
- It is not suitable for detecting mutations at the base pair level, unless that mutation creates or destroys an RE cutting site.

Prehybridization and hybridization solution may contain some or all of the following ingredients.

- **Formamide:** This alters the T_m so the hybridization can proceed at a lower temperature.
- **Salt** (e.g. SSC): Provides ions for the reaction.
- **Denhardt's solution:** Contains the heavy chemical ficoll to weight the solution on to the filter. This solution also contains bovine serum albumin (BSA) which acts as a protein blocker.
- **SDS** (sodium dodecyl sulphate): A detergent which wets the filter evenly and also provides weight.
- **Dextran sulphate:** Speeds up the hybridization.
- **Denatured salmon sperm:** Preblocks human repetitive sequences on the filter.

Alternatively a 'Church–Gilbert' solution may be used, comprising phosphates and sodium dodecyl sulphate but no formamide.

Interpretation of Southern blots

Deletions

A clinical probe is designed to be complementary to or to be tightly linked to a disease gene locus. If that gene (or closely associated DNA) is deleted, the probe will have no target with which to anneal. If a Southern blot usually shows a radioactive band in a normal situation, that band will not appear on the X-ray film.

Gene deletions occur in X-linked Duchenne muscular dystrophy (DMD). Around 60% of affected boys have a deletion which can be detected in diagnostic laboratories by PCR (see multiplex PCR). Historically, however, the disease was first screened using cDNA probes and Southern blotting.

Figure 7.6(a) shows a hypothetical example of some typical band patterns produced by a normal boy, a carrier mother and several patients with different deletions. The DNA would have been digested previously, and the absence of various restriction fragments can be seen. As the dystrophin gene is very large this technique needed many probes to cover the large number of exons, and proved too time consuming.

Figure 7.6 *Detection of deletions, expansions and methylation by Southern blotting*

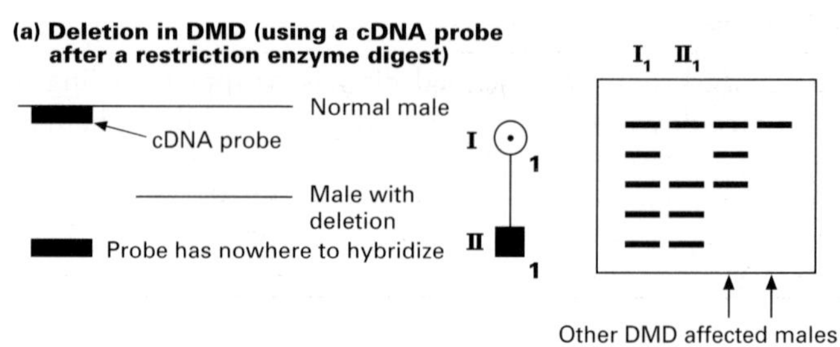

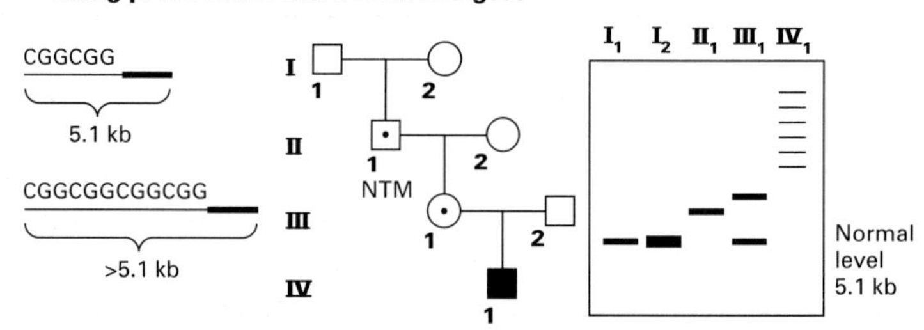

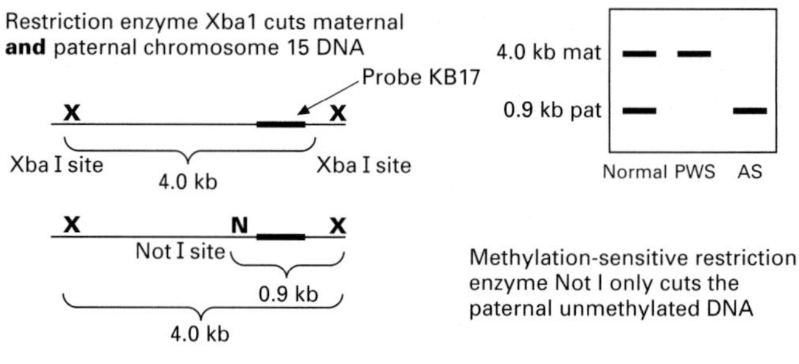

The T_m of a DNA duplex is the melting temperature at which the two strands separate, or alternatively, the temperature at which they begin to hybridize.

For every 1% of formamide concentration, the T_m (in this case the temperature of hybridization) is reduced by 0.7°C.

The advantages of random primed labelling are that the probe need not be pure and only 25–50 ng is required, which is labelled at room temperature to a high specificity (i.e. very radioactive). There is no need to purify the labelled product to eliminate unincorporated nucleotides.

The **method** for achieving **random primed labelling** is as follows.

- Double-stranded probe DNA is denatured by boiling.
- It is added to a solution containing buffer, hexadeoxynucleotides, BSA, water, dATP, dGTP, dTTP, ^{32}PdCTP (the isotope) and the large fragment of the enzyme DNA polymerase 1 (Klenow enzyme).
- The hexadeoxynucleotides or 'hexamers' (which are random sequences of DNA six nucleotides long) find complementary sequences along the single-stranded probe DNA and hybridize to them.
- The Klenow enzyme attaches to the double-stranded template formed by the hexadeoxynucleotides and, using its polymerase function, starts to copy the probe strand, incorporating

Additional DNA

Any increase in the length of DNA in or around a gene will also be detectable using a suitable probe. Occasionally DNA duplications arise, which will increase the size of a characterized restriction fragment. Often, however, duplications are too small to detect by Southern blotting, as are small inserted pieces of DNA.

The major application of Southern blotting has been in the detection of a group of completely different phenotypic disorders known as **triplet repeats** or **dynamic mutations** (see also Chapter 3).

At various locations in the human genome, individuals have runs of three repeating nucleotides, the sequence of which is consistent for a particular disorder. The number of repeats will vary between individuals, and often between the two homologous chromosomes. As a general rule there will be a numerical range of repeats

Table 7.2 *Examples of triplet repeat diseases*

Name of disorder	*Triplet*	*Normal range*	*'Intermediate' range*	*Affected range*
Fragile X syndrome	CGG	0–50 repeats	50–200	>200
Myotonic dystrophy	CTG	5–27 repeats	30–50	50–2000
Huntington disease	CAG	9–27 repeats	27–35	36–121

considered to be 'normal', and another range of repeats considered to be 'abnormal' – usually larger numbers of repeats – which then cause the disease phenotype in affected patients. Some individuals may have an 'intermediate' number of repeats, which can expand into the disease range when passed on to the next generation. These individuals are not usually affected themselves.

Some examples of triplet repeat diseases are given in Table 7.2.

The scenario shown in Figure 7.6(b) represents a typical pedigree from a fragile X family. The mentally normal grandfather II_1 has more than 50 repeats, such that the normal band size detected by the fragile X probe is slightly increased, and does not run so far down the gel. Remember that a male has only one X chromosome and will only produce one band (or allele). This **normal transmitting male** passes on his X to his daughter III_1, and in doing so the size expands by just a few repeats.

She has one normal X from her mother, plus the slightly expanded X from her father; she therefore has two bands of different sizes. She is mentally normal but carries a **premutation.** When she has a son (IV_1), there is a 1/2 chance of him inheriting the expanded X. If he does, the repeat numbers will expand again, and this time produce the disease phenotype.

This **full mutation** can be very unstable when many hundreds or thousands of repeats are involved. It is possible for an affected individual to carry different sizes of repeats in every somatic cell, or to have several clones (groups of cells) comprising large repeats. The clones produce large bands on X-ray film, while the **heterogeneous** (different) populations of cells appear as a smear. This is known as **somatic heterogeneity.**

Females may also carry full mutations (as well as their normal X), and they have an increased risk of being affected (Figure 7.7). As the numbers of repeats increases in both males and females, the fragile X gene becomes more likely to be methylated. It cannot be transcribed, and the normal protein (called FMR-1) cannot be produced. This leads to neurological problems in the brain, resulting in mental retardation.

Methylation digests

This is a specialized form of double digest, where more than one type of restriction enzyme is added to the same patient's DNA. One

the appropriate complementary cold (unlabelled) dNTP or, if the probe sequence contains a G, a 'hot' (radiolabelled) C is added.

- The hot Cs essentially ensure that the whole (copied) probe is radioactive. This reaction proceeds very quickly at room temperature; the incorporation of radioactive isotope reaches a maximum after 2–3 hours.

Hybridization may last from half an hour to 24 hours or more, depending on the type of probe. Hybridization temperature is again determined by the proportion of formamide and is usually between 37°C (with 50% formamide) and 65°C (with no formamide).

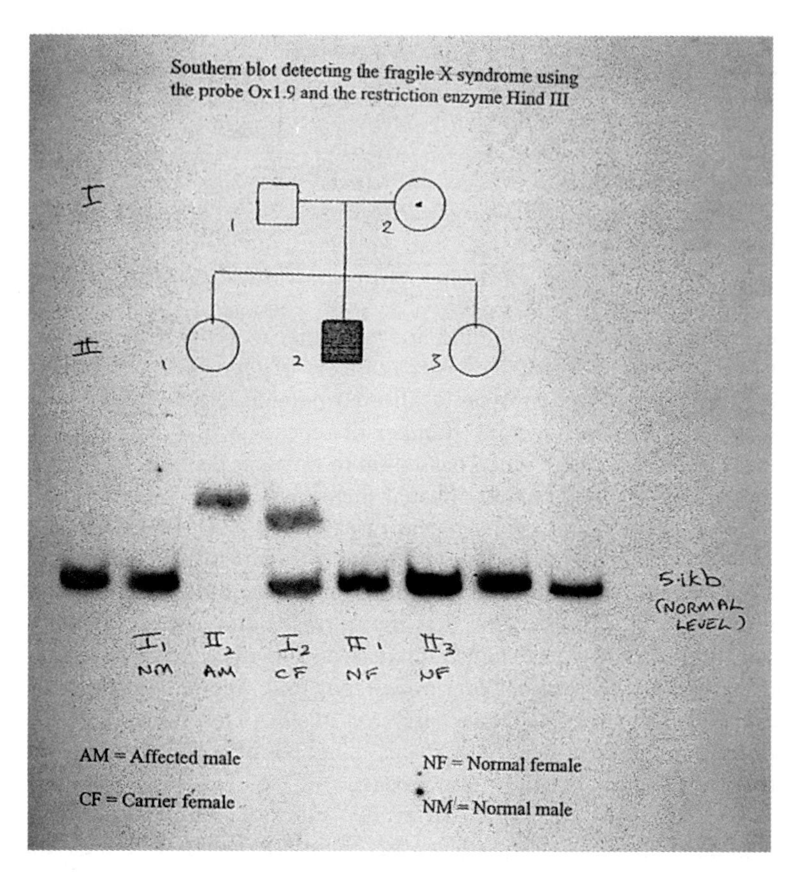

Figure 7.7 *An autoradiograph from a Southern blot of a fragile X family.* I_2 *is the carrier mother,* II_2 *is the affected boy*

of the most common reasons for using this technique is to compare methylated DNA with non-methylated DNA.

Example detecting the parent of origin in Prader–Willi/ Angelman syndrome

We have already studied Prader–Willi and Angelman syndromes by cytogenetic and molecular cytogenetic methods. To study these syndromes at the molecular level, a Southern blot can be done using a methylation digest (Figure 7.6(c)). This exploits the fact that at the PWS and AS region on chromosome 15, the **maternal** DNA is **methylated.** The **paternal** DNA at the locus detected by the probe KB17 is **unmethylated.**

The first enzyme, Xba I (which is **not** methylation sensitive), is added to the patient's DNA, and cuts the DNA from both the maternal and paternal DNA into a 4.2 kb size, detected by the probe KB17.

The second enzyme, Not I (which **is methylation sensitive**), is added; this cannot cut the maternal DNA as it is methylated, but can cut the paternal DNA again, to the smaller size of 0.9 kb.

Following Southern blotting, a normal person will have two bands derived from the two normal parental 15s. A person with Prader–Willi syndrome will only have a 4.2 kb maternal band and a person with Angelman's syndrome will only have the 0.9 kb paternal band.

Although this test does not differentiate between deletion, uniparental disomy or small mutations at the imprinting centre (responsible for the resetting of imprinting of the nearby PWS and AS genes), it will show an abnormal result at the molecular level.

Duchenne's muscular dystrophy was known to be X-linked because of the inheritance pattern. An approximate location was deduced by studying the chromosomes of girls who were phenotypically affected. In each case there was a translocation involving the band Xp21.

The method of **reverse genetics** was used on the globin protein isolated from red blood cells. Once the amino acid sequence was determined, the mRNA sequence could be elucidated. Subsequently mRNA can be converted into cDNA using the enzyme reverse transcriptase. The globin cDNA can then be used as a probe which is complementary to the normal globin sequence.

Probes such as these were useful in diagnosing particular haemoglobinopathies, such as the thalassaemias, where a defective globin protein is produced.

Gene tracking

Background to gene tracking

Historically this was once the most common application of Southern blots. The approximate location of a mutated gene responsible for a specific disorder could often be deduced from gene disruption or position effects resulting from the translocation of a particular chromosome.

The breakpoint would give a starting point from which loosely linked probes could be constructed.

By looking at RFLPs using a variety of restriction enzymes, it could be shown that in some families (with a particular genetic disorder) certain DNA polymorphisms were inherited together with the disease state and the probe was probably linked to the gene locus. Once probes were made which mapped nearer to the gene location (tightly linked), the recombination risk could be reduced and the probe used in clinical diagnosis.

The gene location of disorders such as cystic fibrosis and Huntington disease was discovered by **shotgun cloning**. The first probes were made from known chromosomes, which had been cut into small (probe-sized) random pieces. These are called **gene libraries**.

Many families were then studied who had extensive pedigrees in which the pattern of inheritance was very clear. One of the 'anonymous' probes from a gene library would be used in conjunction with a restriction enzyme known to produce RFLPs. The inheritance of the differently sized RFLPs was compared to the inheritance of the disease.

While it took several years to find linked CF probes, one of the very first probes to be tested for

Examples of gene tracking using linked probes

Using the family pedigrees in Figure 7.8, decide which allele or alleles track with the three diseases, and predict the results of the prenatal diagnoses. (To help you, band sizes have already been transferred to the pedigree.)

Follow rules 1–6 (answers at the end of this section):

(1) The **index case** (the first affected member of a family) must have a **definitive** (clinically definite) **diagnosis**.

(2) The **pattern of inheritance** must be identified (AR, AD or XL).

(3) The band sizes on the autoradiograph must be transferred to the pedigree for each family member. If more than one probe/enzyme result is used, they are each considered separately to begin with.

Huntington disease proved to have a 4% recombination risk, which was low enough to be used in clinical diagnosis.

Care must be taken with heterogeneous disorders. If a probe is used from the X chromosome assuming a diagnosis of DMD and the dystrophy was autosomal, there would be no linkage as the wrong probe was being used.

(4) The band sizes inherited by the **affected index case** are studied **first**. This establishes which size (or sizes) of all the possible choices of alleles is associated with the inheritance of the disease.

(5) The band size(s) associated with the disease are tracked back to the parents, one or both of whom will be carriers depending on the inheritance pattern.

(6) The pattern of the sizes is examined to see how **informative** the probe/enzyme result was.

Tracking in an autosomal dominant disorder: Huntington disease

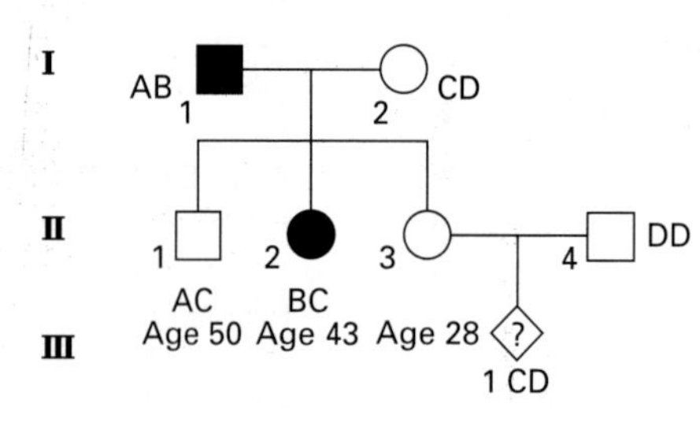

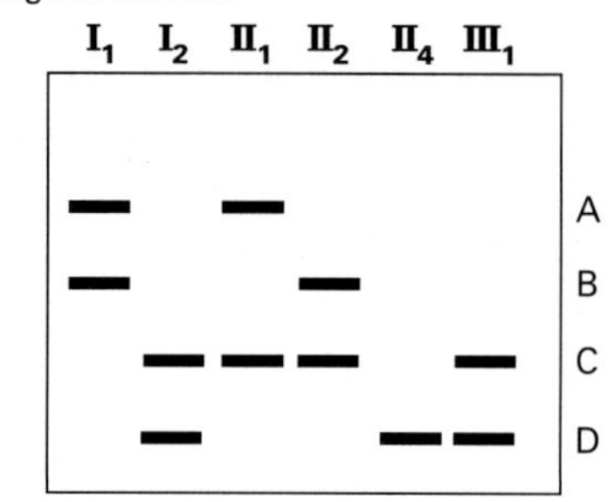

Tracking in an autosomal recessive disorder: cystic fibrosis

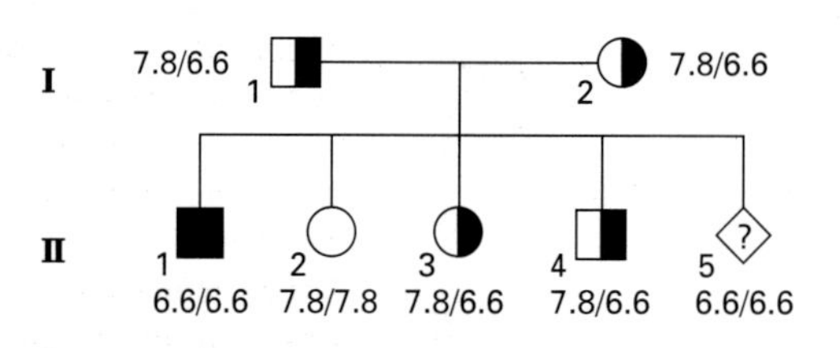

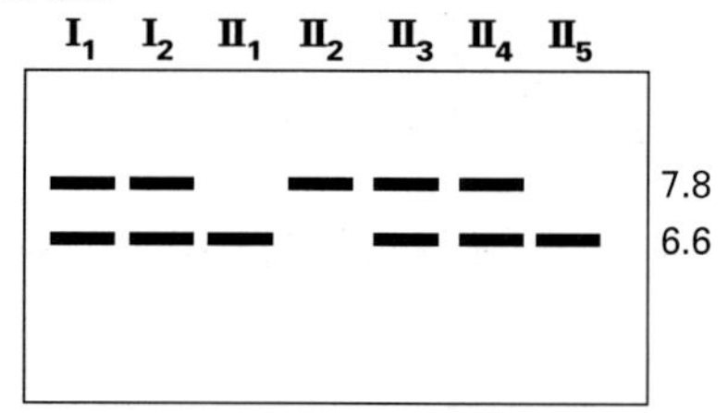

Tracking in an X-linked recessive disorder: haemophilia

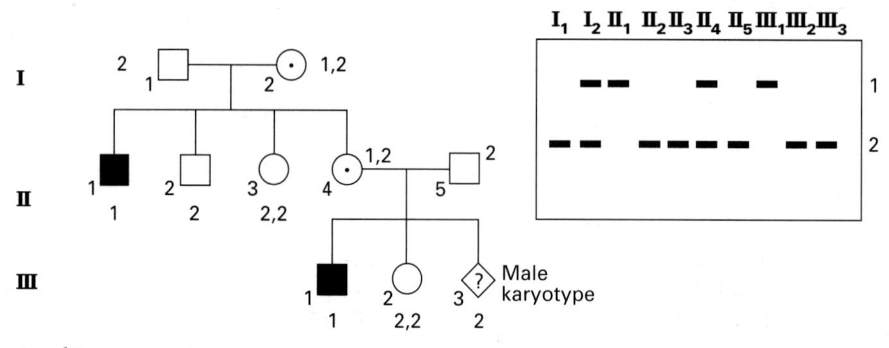

Figure 7.8 *Gene tracking*

Informativeness

RFLPs may produce one of three types of results when used in tracking.

- **Uninformative:** For example, if every member of the family has the same homozygous result. In this instance another probe/enzyme combination has to be tried.
- **Partially informative:** Only limited information may be deduced (for example, if one parent is heterozygous and one parent is homozygous). It may be possible to combine one partially informative probe/enzyme combination with another, such that two partially informative results can be combined as **haplotypes** to give a fully informative pedigree. A haplotype represents all the alleles inherited from one parent on an ideogram of that parental chromosome (Figure 7.9).
- **Fully informative:** Any combination of bands inherited by an individual (e.g. a foetus in a prenatal diagnosis) can be interpreted as either a normal result, a carrier (if recessive) or affected. Often both parents are heterozygous.

With an autosomal disorder there will be a possibility of two sizes of allele corresponding to the two homologues; with X-linked diseases, the female will have a choice of two alleles, but the male has only one X chromosome and will only have one allele.

There are several ways of writing band sizes (the **largest allele** of the pair is usually **written first**):

- The **size** of the bands in kb, e.g. 5.5/4.3 or 2.1/1.4.
- **Letters** of the alphabet, e.g. AB or CD (A > B > C > D).
- **Numbers**, e.g. 1,1 or 1,2 or 2,2 (1 > 2 > 3 > 4).
- + or −, where + represents the presence of a restriction site and − the absence of a restriction site, e.g. +/+ or +/− or −/−. Here, a '+' will be the smaller allele.

Answers to Figure 7.8

- HD = normal. The disease tracks with the B allele. Notice that the status of the parent II_3 need not be known.
- CF = affected. The disease tracks with the 6.6 kb allele from each parent.
- Haemophilia A = normal. The disease tracks with allele 1.

Tracking using RFLPs must be applied each time to a new unrelated family. It cannot be assumed that an allele tracking with the disease state in one family will do so in another (it may track with the normal state the next time). Remember that linked **RFLPs** are **indicators** or **markers** of the disease state, not the disease itself.

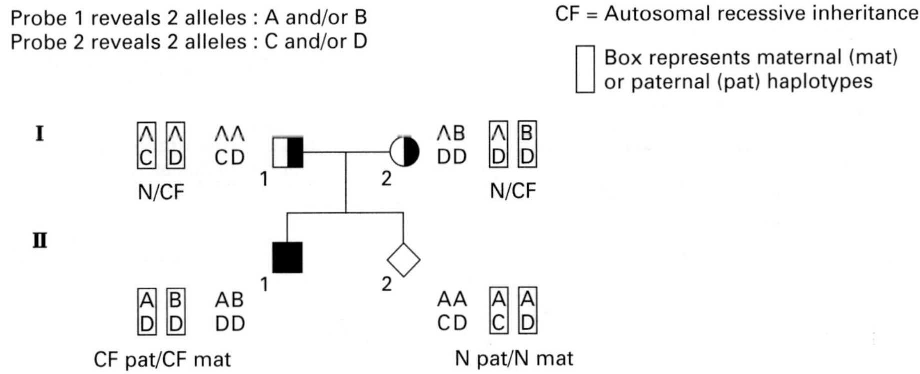

Figure 7.9 *Using haplotypes to produce a fully informative pedigree in cystic fibrosis*

Although theoretically as little as one cell nucleus can be amplified in a PCR, in practice usually 25–500 ng of genomic DNA is used.

Kary Mullis was responsible for the concept of PCR, which initially was done manually and was very labour intensive as the inefficient enzyme Klenow polymerase had to be added at every cycle.

PCR eventually became automated following the use of the **heat-stable taq polymerase**. This is an enzyme isolated from the bacterium *Thermus aquaticus*, which lives in hot springs at 65°C, and was therefore robust enough to withstand the high temperatures and many cycles of a PCR reaction.

The polymerase chain reaction (PCR)

Principles of PCR

PCR is a method of amplifying very small amounts of DNA. Total genomic DNA is denatured and the length of DNA to be amplified is delineated by short pieces of complementary DNA called **primers** which anneal to the target DNA.

A heat stable polymerase attaches to this double-stranded template and copies the existing DNA strands. There is now twice the initial amount of DNA (four strands instead of two). The heat cycles are repeated, ensuring exponential amplification of the target DNA (Figure 7.10).

Details of a typical PCR method

The polymerase enzyme used in PCR limits the size of fragment to be amplified; usually the optimal size is up to a few hundred base pairs. In practice, therefore, exons of genes are suitable targets for amplification.

Primers are designed complementary to **conserved** DNA sequences flanking the chosen exon. The size of the PCR product is determined by the distance between the annealed primers. PCR is

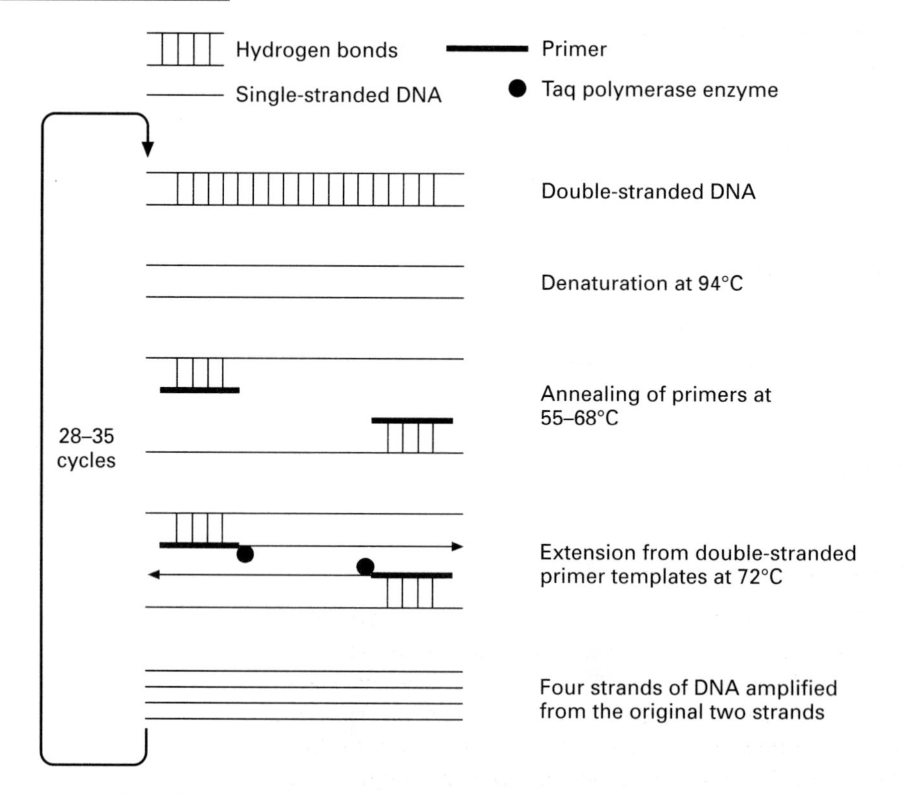

Figure 7.10 *Principles of the polymerase chain reaction*

therefore most often applied to diseases where the gene location is known.

A typical PCR is set up using an eppendorf tube for each patient in a contamination-free area. The reaction usually comprises a suitable buffer, the four dNTPs (dATP, dCTP, dGTP and dTTP), the pair of primers, the patient's DNA and the heat-stable enzyme taq polymerase. The eppendorf tubes (which contain 10–100 µl of the above reaction mixture) are placed in an automated, programmable heating block, which proceeds through several heating cycles as follows.

- **Denaturation** (94°C): The patient's DNA becomes single-stranded.
- **Annealing** (55–66°C): The single-stranded primers anneal to the complementary sequences on the two strands of the patient's DNA.
- **Extension** (72°C): The taq polymerase attaches to the double-stranded template and copies the appropriate strand.

Usually 28–35 cycles are sufficient to produce 10^5–10^6 copies of DNA.

Primers are short, single-stranded pieces of DNA around 18–30 bp in length. If the primer is shorter than 18 bp, the sequence of the target may not be unique in the genome; if the primer is longer than 30 bp, there may be problems with the length of time the primer takes to anneal.

It is hypothesized that Abraham Lincoln had Marfan syndrome, typified by tall thin stature, long fingers and weakness of the aorta (the main artery of the heart). It has been suggested that a PCR could be done from the dried blood of the jacket he was wearing when he was assassinated in order to study the fibrillin gene which is mutated in Marfan syndrome.

Advantages of PCR

- Very small amounts of DNA are required.
- The procedure is fast; results are obtainable in hours (3–48 hours, depending on the steps following PCR).
- PCR is safe; no radioactivity need be used.
- Mutations can be detected at the molecular level.

Disadvantages of PCR

- Large alleles cannot be amplified.
- Some of the DNA sequence usually has to be known to design the primers.
- The reaction is easily contaminated.
- Dosage is difficult to estimate.

Applications of PCR

Apart from clinical diagnostic use, PCR may be applied to such areas as **forensic science** or **archaeology** and **anthropology** (and hence evolution), where only minute traces of DNA may be found. If primers are used which can detect numerous polymorphisms, PCR can be used in **paternity testing** and also criminal pathology, as the polymorphic band patterns will be almost unique to an individual, and constitute a **DNA fingerprint**. The more closely two

RT-PCR (reverse transcriptase PCR) uses **RNA as a starting point** rather than DNA. RNA is a short-lived molecule which may only exist in small amounts and only when a gene is actively **expressing** a product. The reaction proceeds in two stages: an enzyme called reverse transcriptase 'transcribes' the mRNA into cDNA, which is then amplified in the usual manner. A typical application would be in the study of **leukaemias**, where the reappearance of particular abnormal clones may be detected very efficiently.

As CF is an autosomal recessive disease, the affected children have two mutations (which may be different), one on each chromosome 7. The mutations may result in an altered or truncated protein. The normal protein is called CFTR (cystic fibrosis transmembrane conductance regulator); its normal components form a chloride channel which is essentially a pore in the epithelial cell membrane. In the pathological state, chloride ions and hence sodium ions are not reabsorbed in the appropriate area of the sweat gland due to the absence of a chloride channel. This leads to an increased salt concentration in the sweat.

When a carrier produces two different sizes of PCR product, they attempt to pair, but due to the size discrepancy a unique band pattern is produced which runs at a different level to that expected on the gel. This is a **heteroduplex**, which has a different **conformation** to the correctly paired 98 bp or 95 bp products.

people are related, however, the greater the likelihood of them inheriting the same polymorphisms. Identical twins will share identical genetic fingerprints.

Interpretation of PCRs

Direct visualization of PCR products

If a pair of primers flank a region that displays size variations, then the products themselves will also vary in size. This is useful where a known deletion, insertion or expansion is characteristic of a particular disorder.

The cystic fibrosis mutation ΔF508

The most common cystic fibrosis mutation in the Caucasian population is ΔF508 (Δ = deletion, F = phenylalanine at amino acid position 508) in exon 10 of the CF gene. This is found in 70–80% of affected children.

The **deletion** of 3 bp (CTT) can be detected by running a high percentage polyacrylamide gel to separate the differently sized PCR products produced by normal individuals, carriers and affected patients (Figure 7.11).

A normal product using a specific pair of primers is 98 bp in size; the deleted product will only be 95 bp. A carrier will therefore display one 98 bp (normal) product from one chromosome 7 **and** a 95 bp (deleted) product from the other chromosome 7. An affected child will have two faulty chromosomes; the gel will display two 95 bp products (indistinguishable on the gel). The gel is stained with ethidium bromide for visualization (Figure 7.12).

The triplet repeat of Huntington disease

Primers have been designed to flank the region of CAG repeats in HD which may expand in some families, resulting in affected individuals.

When run out on a gel with suitable molecular weight markers giving bands of known sizes, an estimate of the numbers of repeats can be obtained. It can then be determined whether a patient is unaffected (up to 34 repeats), affected (over 36 repeats) or borderline (34–36 repeats). In the last category it is difficult to assess whether these numbers of repeats may expand in future generations.

The duplication in hereditary motor sensory neuropathy (HMSN)

The hereditary motor sensory neuropathies are a heterogeneous group of neurological disorders also known as Charcot–Marie–Tooth disease. The most common form is inherited in an autosomal

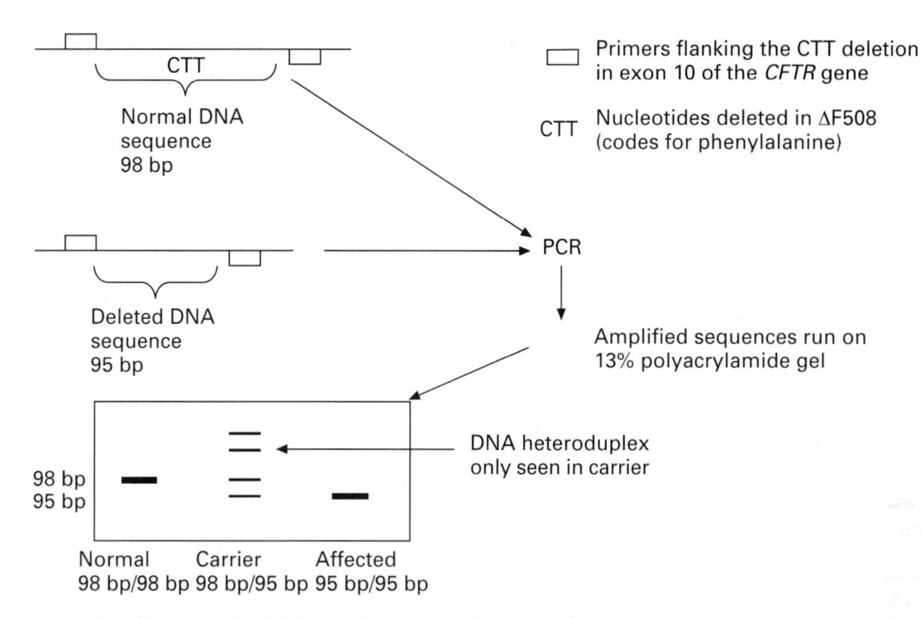

Figure 7.11 *Direct visualization of a PCR product using the cystic fibrosis mutation ΔF508*

dominant manner, and is caused by a duplication of part of the short arm of chromosome 17.

The duplication can be detected in the laboratory by using different pairs of PCR primers (or markers) which map to the duplicated region. In the normal state two alleles will be detected. If there is a duplication, either three alleles are detected or there is a **dosage** effect such that two alleles are detected, one of which may be visible as a heavier band. One of the disadvantages of PCR becomes apparent when looking at 'dosage'; it takes experience to

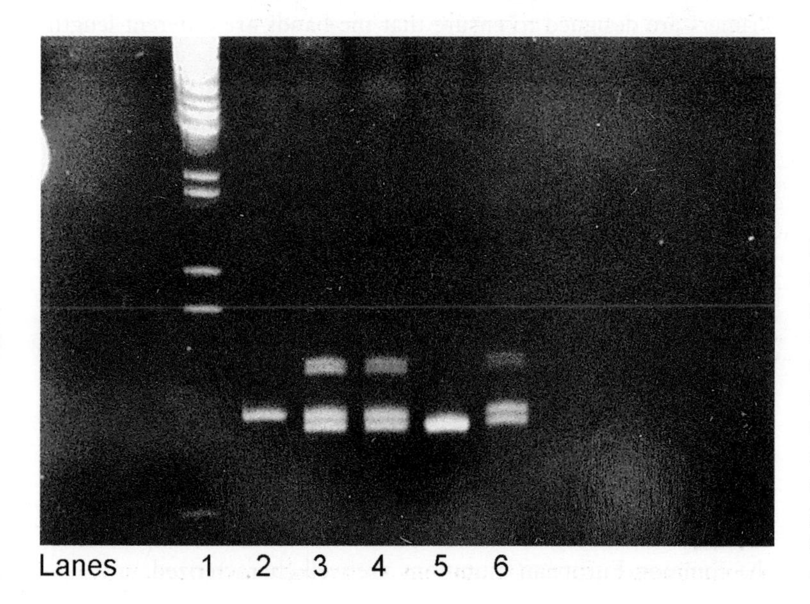

Figure 7.12 *A polaroid of a gel showing detection of the ΔF508 mutation. The bands of DNA are stained with ethidium bromide. Lane 1 is a molecular weight marker, lane 2 is a normal child (N/N), lanes 3 and 4 are the heterozygous carrier parents (N/ΔF508), lane 5 is the affected index case (ΔF508/ΔF508) and lane 6 is a prenatal diagnosis revealing the foetus to be a phenotypically normal carrier of ΔF508 (like the parents)*

During the PCR of the **exon 11 CF mutation G551D** primers which flank exon 11 of CF amplify a product of 425 bp in every individual, and thus a further step is necessary to detect the mutation.

The mutation G551D is a G to A base substitution such that glycine (**G**) at amino acid position **551** is changed to aspartic acid (**D**). Fortuitously, a restriction site for the restriction enzyme Mbo 1 is created in individuals carrying the mutation when the G to A transition occurs, which cuts the 425 bp product into two fragments: one of 242 bp and one of 183 bp. In normal individuals the restriction site is not present and Mbo I can no longer cut, resulting in one fragment of 425 bp.

Carriers will have one normal chromosome (which does not cut) and one chromosome with the mutation (which does cut) – so they will have three bands.

A typical example of **mutation nomenclature** is given by the CF mutation G542X:

- **G** is the amino acid glycine coded for by the **normal DNA sequence**;
- **542** is the **amino acid position** in the protein;
- **X** is a stop codon coded for by the **mutated DNA sequence**.

The first letter is therefore the normal amino acid, the number is the amino acid position, and the last letter is the changed amino acid.

know when more than one copy of an identical allele is present, and often there is preferential amplification of the smaller allele. Automated techniques are now used to standardize dosage analysis.

Restriction enzyme digests and PCR

A restriction enzyme digest may be performed on the PCR product from a patient, exploiting the fact that a mutation may create or destroy a restriction site.

Amplification refractory mutation system (ARMS)

A pair of primers is designed such that one primer is complementary to a common conserved sequence, both in normal individuals and in patients carrying a particular mutation. The other **primer** is only **complementary to the mutated sequence.**

Only in DNA carrying the mutated sequence will PCR amplification succeed, as **both** primers anneal, enabling visualization of a band. In normal individuals, only **one** common primer will anneal; the other primer is not complementary to the normal sequence. No band will be produced.

Multiplex PCR

Rather than amplify one exon of a gene individually using one pair of primers, it is more efficient to use **more than one pair of primers** (for multiple exons of a gene) in a reaction tube.

Several bands will result, each representing different exons. Primers are designed to ensure that the bands are different lengths and are well separated on the gel.

Problems may arise if too many sets of primers are used which do not respond equally to a common PCR buffer – some exons may 'drop out'.

ARMS multiplex

If a particular disorder can be caused by several different mutations, each expressing at a reasonably high frequency (say $>1\%$), one of a pair of primers can be designed complementary to one particular mutation. This is done for a number of common mutations (**ARMS**), and all the pairs of primers are put into one PCR reaction tube (**multiplex**), together with a pair of control primers which detect PCR failure. This has proved to be an efficient method of screening **cystic fibrosis** mutations, where frequencies of the common European mutations are well characterized.

Screening strategies using PCR

The following techniques are useful as they are able to detect differences in band positions on gels (representing mutated DNA sequences) with high sensitivity. However, the exact nature of the mutation is not usually known, and the position of the mutation may only be traceable to a particular exon determined by the PCR.

Heteroduplex analysis (HA)

HA is especially useful for carrier detection, as carriers have one normal and one mutated DNA sequence.

PCR products are denatured and slowly allowed to reanneal at room temperature such that the mismatched normal and mutant strands display a different band pattern on the gel compared to homozygous sequences. If detection of mutant homozygotes is required, the suspected mutant DNA must be mixed with an equal quantity of normal DNA.

Single-stranded conformational polymorphism (SSCP)

PCR products are denatured and run as single DNA strands such that each strand assumes a particular single-stranded conformation. A normal sequence strand will differ from a mutant strand and produce a different band pattern.

Denaturing gradient gel electrophoresis (DGGE)

DGGE is more sensitive than HA or SSCP, but needs specialized apparatus. Double-stranded PCR products are run on gels made up of increasing amounts of chemical denaturant at 65°C. When the double strands denature, the DNA is retarded at a certain point in the gel. A carrier (whose normal and mutant sequences may differ by only one base pair) will show band patterns representing both **normal and mutant sequences, which will have separated** at **different positions on the gel.**

Sequencing

Once a mutation has been detected by a primary screen, its nature and position have to be characterized by determining the sequence of nucleotides on one DNA strand. Older chemical methods such as the Maxam–Gilbert technique have been largely superseded by enzymatic methods such as the **Sanger dideoxynucleotide** technique.

A sequencing **primer** (yyy) is normally designed complementary to a **known sequence** (xxx) at the 3′ end of the single DNA strand to be sequenced. Four tubes are prepared; in each tube are three

In order to differentiate the carriers of a specific mutation from homozygous affected patients, a primer complementary to the non-mutated DNA sequence can be run with the common primer in a separate ARMS reaction.

The **Duchenne muscular dystrophy (DMD)** gene lends itself to the **multiplex PCR** especially well. As the gene is 2.4 Mb and about 78 exons long, examining every patient exon by exon for deletions in the DMD gene would be incredibly time consuming. For this reason up to nine pairs of primers may be used simultaneously, representing two key areas of the gene (the 5′ end and the 3′ end) which are deletion hotspots. A multiplex approach enables 99% of deletions to be detected.

Although primer pairs are designed to be sequence specific, occasionally the primer match is not perfect, and spurious bands may result. A few microlitres of the first PCR product can be taken and used as the DNA source in a second PCR. **Nested PCR primers** may be designed, both of which are located internally (on the DNA sequence) to the initial pair of primers.

This technique can be used in preimplantation diagnosis, which uses a single embryonic cell as the initial DNA source.

Modifications of the HA technique include chemical and enzymatic cleavage of mismatches. At the point where the heteroduplex does not match due to a mutation, either chemicals such as osmium tetroxide, or bacteriophage enzymes can be used to cleave one strand of the heteroduplex at the mutation site, producing differently sized fragments on a gel.

In a dideoxynucleotide, the OH^- group on the 3′ carbon of the sugar ring has been replaced by a hydrogen. A further nucleotide cannot be added as no phosphodiester bond can be formed.

deoxynucleotides (dNTPs), with the fourth containing a mixture of deoxynucleotide and dideoxynucleotide (ddNTP):

Tube 1:	dATP	dCTP	dGTP	dTTP	**ddTTP**
Tube 2:	dATP	dCTP	dGTP	**ddGTP**	dTTP
Tube 3:	dATP	dCTP	**ddCTP**	dGTP	dTTP
Tube 4:	dATP	**ddATP**	dCTP	dGTP	dTTP

If we take the hypothetical sequence 5′GATCCATxxx3′, the primer (yyy) binds to the complementary 3′ end. This enables a DNA polymerase (which is added to the four separate tubes) to start copying the sequence at the double-stranded template, incorporating the complementary nucleotides. This results in a complementary strand which will be sequenced in a 5′ to 3′ direction:

DNA strand to be sequenced: **5′GATCCATxxx3′**
Complementary strand: **3′CTAGGTAyyy5′**

However, because of the ratio of deoxynucleotides to dideoxynucleotides, every time the enzyme tries to incorporate a dideoxynucleotide the reaction is stopped, and a shorter piece of DNA results.

For example, the first complementary nucleotide should be an 'A' (the corresponding dideoxynucleotide is in tube 4). By chance, the reaction may truncate at position 1 if the dideoxynucleotide version of 'A' is incorporated. If not, the reaction proceeds to nucleotide 'A' at position 5 instead, producing a longer piece of DNA.

We now know at which positions 'A' appears in the sequence – position 1 and position 5. Taking all four tubes into account, all possible lengths of DNA are generated (Figure 7.13).

These fragments are now run on a long 6% denaturing gel (using urea as the denaturing agent), allowing hundreds of base pairs to be sequenced. The four reactions are run in separate wells, and the sequence of bands will appear as seen in Figure 7.13.

Using computer software, sequencing has now become an automated procedure, revealing any changes in bases from those seen in a normal DNA fragment.

While this may be seen as the ultimate detection system, it must be remembered that sequencing generates a vast amount of information, and is not often used as a primary screen. Sequencing is more often used to characterize the exact nature of a change in a DNA sequence which has been detected in a primary screen by one of the methods mentioned previously, and may confirm that a pathological mutation is present rather than a harmless polymorphism.

Summary

Since the 1970s molecular genetics has become the tool of the geneticist studying human disease. From the earliest Southern blots

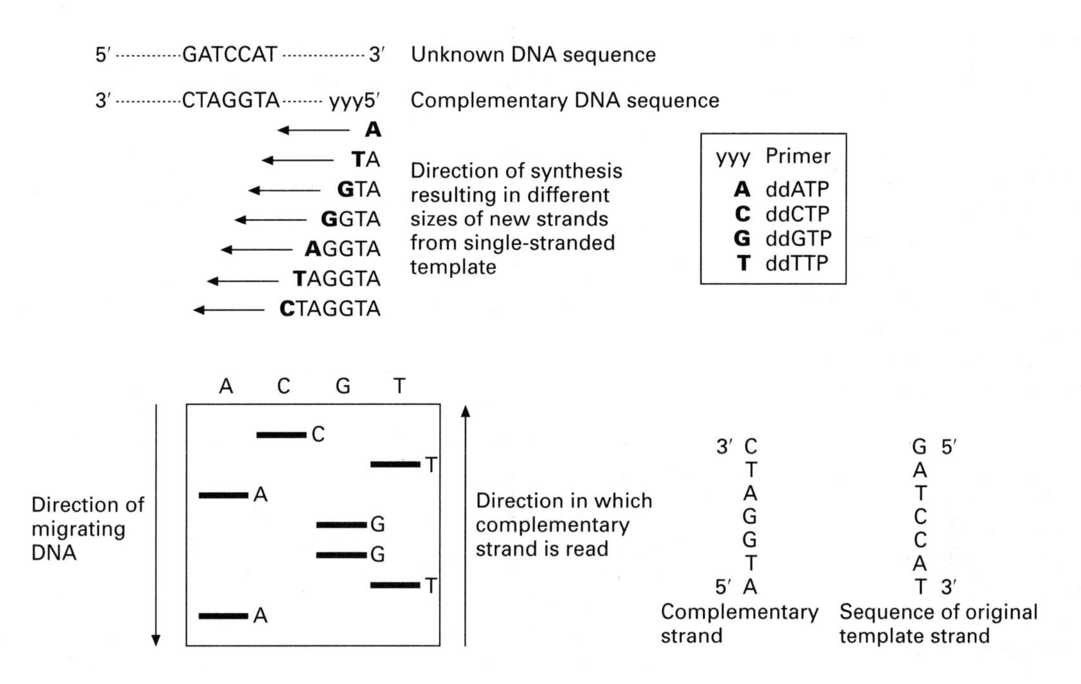

Figure 7.13 *DNA sequencing*

using loosely linked probes through to the polymerase chain reaction used for mutation analysis in clinical genetics and for fingerprinting in forensic science, sophisticated screening and sequencing techniques have been developed which are used in large laboratories today.

Suggested further reading

Clark, D.P. and Russell, L.D. (1997). *Molecular Biology Made Simple and Fun*. Cache River Press.

Strachan, T. and Read, A.P. (1999). *Human Molecular Genetics 2*. Bios Scientific Publishers Ltd.

Self-assessment questions

1. Name three tissues from which DNA may be extracted. What is the defining factor with respect to the cells which ensures the presence of DNA?
2. Two fragments of DNA must be separated by agarose gel electrophoresis. Their sizes are 5.2 kb and 2.8 kb. What percentage gel might you use and why?
3. Part of an intronic double-stranded DNA sequence runs as follows:

5′AGCCTCGAATTCATCGAG3′

A mutation occurs such that the sequence now runs:

5′AGCCACGAATTCATCGAG3′

How many pieces of DNA would you get if you digested with the restriction enzyme Taq I for each sequence? Would the mutation have a clinical effect on the person? What name is given to these differences in sequences?

4. Using all the DNA fragments from Question 3, in which order would they run on a gel (use bp)? Now using the ABCD system, allocate letters to each fragment.
5. Which DNA technique detects large pieces of DNA? Give an example of a disorder to which this applies.
6. What is a linked probe? What disadvantages and advantages are there to using a linked probe?
7. What is the major advantage of the PCR technique compared with Southern blotting? Why is it useful to know the sequence of the gene or exon of interest?

Key Concepts and Facts

Molecular Genetics

- Technology is used which works at the DNA level and therefore makes it possible to detect point mutations and other small changes undetectable by conventional cytogenetics and FISH.

Probes

- Probes used in routine clinical work are usually pieces of double-stranded DNA, either complementary or linked to a region of interest. Probes are inserted into vectors for cloning.

Polymorphisms

- Changes in DNA sequence may either lead to a clinical defect or a harmless non-clinical change termed a polymorphism. These can be revealed as different lengths of DNA using restriction enzymes.

Southern Blots

- These are used when large pieces of DNA are to be detected, or when the gene locus is not known. An appropriately labelled probe is hybridized to the patient's denatured DNA and the result detected as a band pattern.

Gene Tracking

- This is an indirect method of predicting disease or carrier outcome where the gene locus is not known. A closely linked probe detects polymorphisms, one of which is inherited (and therefore tracks) along with the disease status in a family.

PCR

- A method of producing up to a million copies of DNA from very little starting material. It is quicker and safer than Southern blotting, although the conserved sequences flanking the exons of a gene usually have to be known, and there is a contamination risk. PCR is used in forensic science, archaeology and studies of evolution, as well as for clinical diagnosis.

Mutation Screening

- If a disease is characterized by many different mutations, it is easier to use a pre-screening method such as HA, SSCP or DGGE to look for general changes in DNA conformation. Specific mutation analysis or sequencing can then be used to determine the exact nature and position of the mutation.

Chapter 8
Cancer genetics

Learning objectives

After studying this chapter you should confidently be able to:

State the difference between a constitutional and a somatic mutation.

Describe the difference between inherited and acquired cancer mutations.

List examples of environmental factors in cancer.

State the normal function of proto-oncogenes, tumour suppressor genes and mismatch repair genes.

Outline how mutations in the above genes lead to loss of cell control, and list examples.

Describe the role of p53 in the cell cycle and apoptosis.

Define Knudson's two hit hypothesis.

Describe the role of chromosome analysis in the identification and prognosis of cancer.

Outline the multistep nature of cancer together with the limitations of this concept.

In 1890 David Hanseman suggested that there was a connection between the abnormalities he was finding in tumour cell nuclei and the origin of cancer.

In 1914, Theodor Boveri published his hypothesis which stated that cancer does have a genetic component such that chromosomal aberrations were responsible for the origin and malignant progression of cancer.

Cancer usually develops in a series of steps. There may be an inborn **genetic** predisposition due to an abnormality inherited via the germ cells (a **constitutional** mutation), or an abnormality **acquired** during the lifetime of an individual (a **sporadic** mutation in a small number of somatic cells).

Sometimes an environmental component contributes to or causes the genetic mutations leading to the development of cancer.

A **mutagen** is an agent that can cause mutations in DNA and a **carcinogen** is any agent that may induce cancer.

Examples are chemicals (such as benzene or the tars in cigarettes), radiation and viruses.

There are three main classes of genes which, when mutated, may lead to cancer. These are:

- **proto-oncogenes**
- **tumour suppressor (TS) genes**
- **mismatch repair genes.**

Between 1910 and 1914 Professor Peyton Rous showed that the lysate (clear liquid) from the tumour cells of chickens could be injected into other chickens, who subsequently developed tumours. The virus isolated from the lysate was called **R**ous **s**arcoma **v**irus (RSV).

Oncogenes

There are very few viruses in humans known to produce tumours. Human oncogenes were discovered through the study of animal tumour viruses, in which oncogenes were first identified.

Viruses either have a DNA or an RNA genome; the animal tumour viruses most closely studied have an RNA genome and are called **retroviruses.**

- As part of their life cycle retroviruses use the enzyme reverse transcriptase to make a double-stranded DNA copy of their RNA. This integrates into the host genome; the virus then uses the host cell's metabolism to produce new viral proteins.
- Some viruses have a gene that can induce tumours in animals. This is the **viral oncogene** (v-*onc*).
- Viral oncogenes may arise following errors in viral replication subsequent to integration into the host. When the virus is excised it takes the host gene with it.
- Animal oncogenes were first discovered due to the remarkable similarities between the viral oncogenic sequences and parts of the animal genome. This reflects the animal origin of the viral oncogene.
- Although there are very few examples of human retroviruses, animal examples such as RSV enabled the characterization of human oncogenes.

DNA viruses, unlike retroviruses, can replicate autonomously without being integrated into the host genome.

Retroviruses that do not carry an oncogene can inappropriately modify a host gene upon integration, or may modify the expression of the host gene using a powerful promoter.

Examples of human tumour or cancer viruses include the Epstein–Barr virus and the papilloma virus (both with DNA genomes), and the retrovirus HTLV-1, which has an RNA genome and is responsible for adult T-cell leukaemia-lymphoma (ATLL).

Function of proto-oncogenes

Certain human genes code for various proteins controlling growth and differentiation. Activating these at the correct time and in the correct tissue results in normal cell proliferation.

Growth genes displaying their **normal** functions are called **proto-oncogenes.** They have the potential to be activated at the wrong time or in the wrong place; they would mainly exhibit a gain of function and become **cellular oncogenes.** By studying what happens to cells when a proto-oncogene is transformed into a cellular oncogene, we can deduce the normal function of that gene.

The activation of an oncogene generally results in a **loss of control of cellular proliferation and differentiation.** This implies

As viruses can pick up these human or animal cellular genes, a nomenclature has developed using a prefix to differentiate between cellular and viral oncogenes. The oncogenes themselves are usually represented by three letters which reflect the tumour origin. Examples are:

- *MYC** = avian **myelocytoma**-tosis
- v-*onc* = **v**iral oncogene
- c-*onc* = homologous **c**ellular proto-oncogene

The distinction between **proto-oncogenes** (the c-*onc* genes) and the term **oncogene** is gradually becoming outmoded; the term oncogene is often used for the normal gene and the term activated oncogene is used for the abnormal version (Strachan and Read, 1999).

* According to the 1987 Guidelines for human gene nomenclature, human genes are written in italicized capitals and their gene products are written in non-italicized capitals (see also Chapter 2).

Table 8.1 *Classes of oncogenes*

Group	*Comments*	*Example*
Growth factors	The viral oncogene v-*sis* is homologous to part of the human growth factor gene *PDGFB*	*SIS*
Growth factor receptors	These cell surface receptors receive signals from the growth factors	*ERBB*
Signal transduction systems	Involved in intracellular signalling	*HRAS*
Nuclear proteins	Bind DNA (e.g. transcription factors)	*MYCN*
Cell cycle related	Play a normal role in progression through the cell cycle (e.g. cyclins)	*MDM2*

that the normal role of proto-oncogenes often involves the cell cycle. The finding that proto-oncogenes may code for growth factors and their receptors (see Table 8.1 for more details) supports this.

Oncogenes tend to be grouped into classes based on either the normal genes they code for or where their proteins are found in the cell.

How oncogenes exert their effect

Oncogenes exert their effects in two main ways (Figure 8.1):

- **Increasing the amount of normal protein.**
- **Production of an altered protein.**

A proto-oncogene may be induced to produce an increased amount

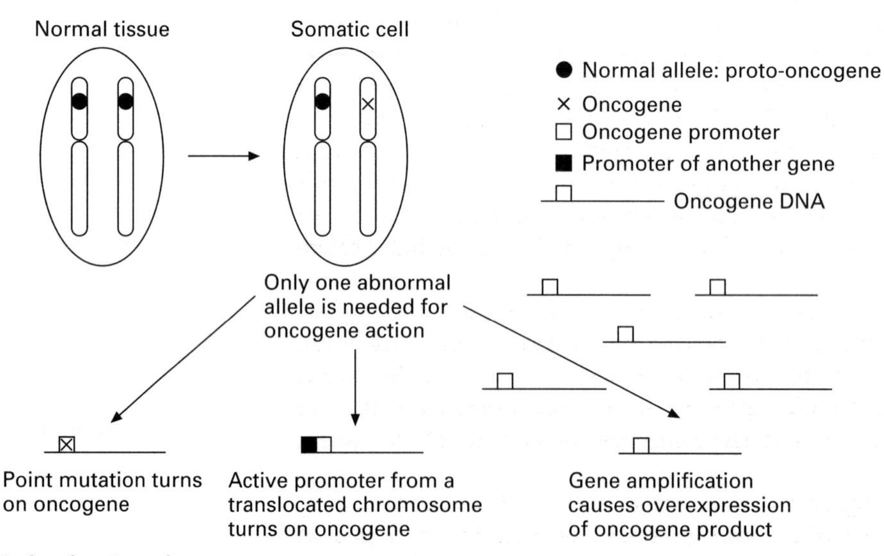

Figure 8.1 *Mode of action of oncogenes*

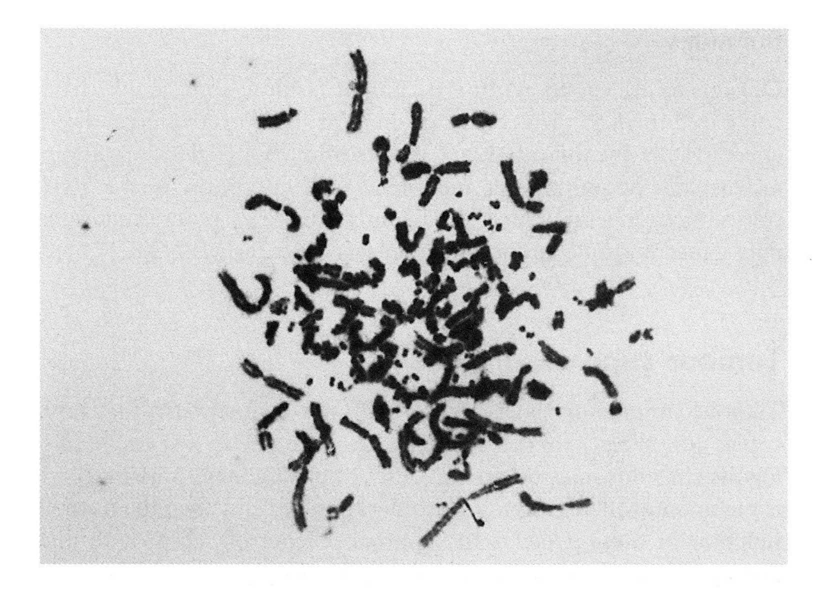

Figure 8.2 *Chromosomes from a metaphase cell of a neuroblastoma showing double minutes*

of product. This may also be in the wrong cell type. The result will be **uncontrolled overexpression.**

Proto-oncogenes may be amplified such that there are tens or hundreds of copies of the resultant cellular oncogene in the cell nucleus. In the case of the childhood thoracic/abdominal tumour neuroblastoma, the amplification of the oncogene *MYCN* (on chromosome 2q) is sometimes cytogenetically visible in chromosome preparations as tiny paired spots of DNA called double minutes (Figure 8.2). The amplified *MYCN* can also appear as non-banding pieces of translocated DNA called HSRs (homogeneously staining regions).

The *MYC* gene may be amplified in response to exposure to methotrexate – a change which confers resistance to this chemical.

Normally **inactive proto-oncogenes can be activated** by moving them next to an active promoter. This happens in Burkitt's lymphoma, where the chromosomes may display three types of translocation. The most common is the t(8;14), but t(2;8) and t(8;22) are also seen. The *MYC* oncogene on chromosome 8 is translocated next to the active promoter of the immunoglobulin heavy chain gene on chromosome 14, and is subsequently activated itself. The loci on chromosomes 2 and 22 are genes coding for the immunoglobulin kappa and lambda light chains respectively.

Gene activation may also occur through the creation of a hybrid or chimeric gene, caused by one translocated gene fusing with another. An example is seen in chronic myeloid leukaemia (CML), where the *ABL* gene on 9q34 is reciprocally translocated to chromosome 22, and fuses with the breakpoint cluster region (*BCR*) at 22q11.

A proto-oncogene may be turned into a cellular oncogene by mutations in the DNA. Point mutations have been found in the *RAS* oncogene for example.

Summary

Oncogenes appear to act in a dominant fashion; some books have compared them to the accelerator on a car – once activated, a direct effect results in abnormal cell proliferation. Oncogene mutations are usually acquired, not inherited. One exception is the *RET* proto-oncogene, in which dominantly inherited point mutations may cause multiple endocrine neoplasia type 2a (MEN2a).

Tumour suppressor genes

Tumour suppressor genes act in a recessive manner such that **both** copies (or alleles) of the normal gene have to be lost on homologous chromosomes before the onset of malignancy. If malignancy is the uncontrolled growth or proliferation of cells, then the normal function of these genes is to suppress or control cell growth and progression.

The existence of tumour suppressor (**TS**) genes can be revealed in two ways. The first uses hybrid cell lines, whereby normal cells are mixed with cells from a malignant cell line; the hybrid line resumes normal control of cell growth, and the excess growth of the tumour cell line is suppressed. This implies that normal gene alleles can override absent or abnormal TS function.

The point at which uncontrolled growth occurs can sometimes be related to the loss of whole or parts of certain chromosomes. Again, this infers that TS genes were present on a particular chromosome and have now been lost, resulting in uncontrolled cell proliferation.

Knudson's two hit hypothesis

In 1971 Alfred Knudson statistically analysed children with the eye tumour **retinoblastoma**, and provided the first practical example of a TS gene using the theory he had formulated from his epidemiological data.

Retinoblastoma is a childhood tumour of the eye that usually appears before the age of 3 years. It can be found in one eye (**unilateral**) or both eyes (**bilateral**), and comprises a mass of undifferentiated retinal cells. Knudson compared the incidence of unilateral or bilateral tumours in patients with or without a family history, and also with the age of onset.

In patients with a family history (around 40%), the tumour often arose in both eyes and had an earlier age on onset. In patients with no family history (around 60%), the tumour was more likely to be unilateral and have a later age of onset.

Knudson suggested that the disease occurred as a result of two consecutive mutations arising in two different ways in families with a history (i.e. hereditary) and without a history (sporadic).

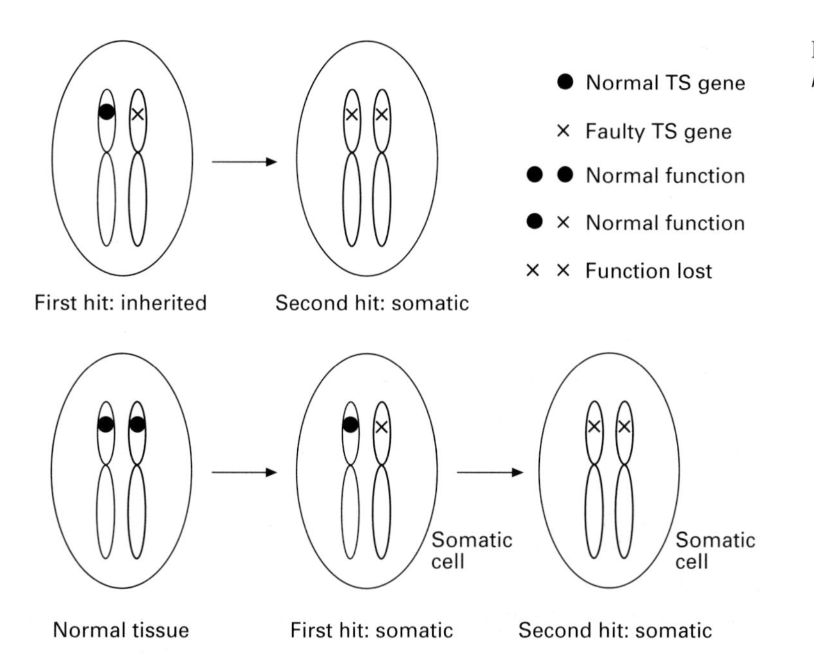

Figure 8.3 *Knudson's two hit hypothesis*

> The presence of a constitutional mutation (the first hit) predisposes a cell to a second somatic mutation (the second hit), therefore the age of onset of retinoblastoma is earlier and the eye tumours are more likely to be bilateral.

> A good reason for not letting young children get sunburned early in life is because their skin cells may receive their first 'hit' at this time from the sun's ultraviolet light, which damages their DNA. This increases their chance of a second 'hit' later in life if they are often exposed to the sun, and may set in motion the malignant changes found in, for example, melanomas.

> The retinoblastoma gene is involved with regulation of the cell cycle. If pRB is unphosphorylated, it can bind another factor E2F, suppressing progression through the cell cycle. If the pRB protein is phosphorylated, it releases the E2F transcription factor which causes the cell to cross the G1 checkpoint and enter S, hence promoting cell growth and division.
>
> *RB1* is therefore said to connect the cell cycle clock with the transcriptional machinery.

Deletions or mutations of the retinoblastoma gene (*RB1*) lead to absence of the retinoblastoma protein (pRB). This can happen in one of two ways:

- In the familial form, the first mutation is already present in the retinoblasts, having been **inherited** from the parental germ line. The mutation is therefore **constitutional.** The second mutation is **somatic** and occurs in a retinal cell already bearing a constitutional mutation, and therefore **both alleles are lost** (Figure 8.3).
- In the later onset **sporadic** unilateral tumours, **both mutations** (or hits) **arise somatically**; the chance of two allelic mutations eventually occurring in the same cell is low and therefore rare. The probability of a second hit is higher than the probability of the first hit, however, as some of the mechanisms leading to the second hit depend on the existence of the first hit (see the section on LOH).

The two hit theory was later shown to be true when microscopically visible deletions on the long arm of chromosome 13 (13q14) were found in retinoblastoma patients. Non-visible deletions and mutations were eventually confirmed by molecular means, and the retinoblastoma gene is now fully characterized.

Loss of heterozygosity (LOH)

By comparing the patient's somatic cells with their tumour cells, there should be a difference at the molecular level such that there is no working copy of *RB1* remaining in the tumour cells. This may

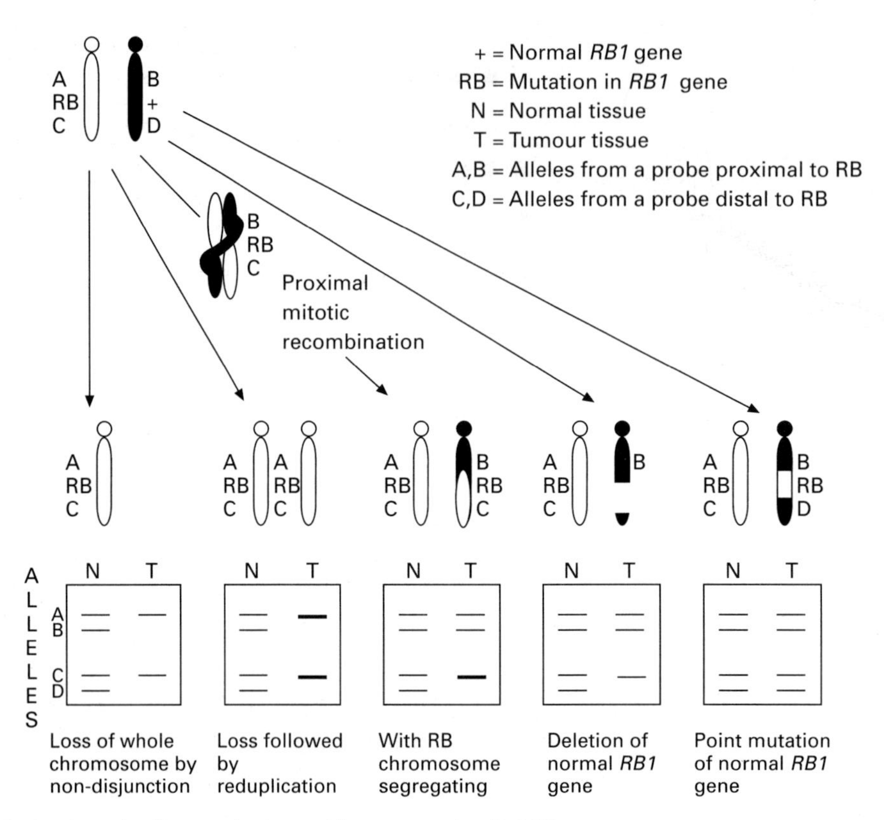

Figure 8.4 *Mechanisms leading to the loss of heterozygosity (LOH)*

A Wilms tumour is an embryonal kidney tumour of children. Knudson's two hit theory can also be applied here, but the sporadic tumours comprise 99% of the total compared with 1% of constitutional tumours. There is an association between Wilms tumours and a deletion of the short arm of chromosome 11 (11p13), and LOH can often be demonstrated. However, there are other Wilms' tumour loci, so that the 11p13 locus is usually designated *WT1*.

The correct way to pronounce apoptosis is with a silent second 'p' so that it sounds like 'apotosis'.

arise by various mechanisms (Figure 8.4), but is sometimes detectable using molecular DNA markers.

It can be shown, for example, that an individual may be heterozygous for a certain polymorphic molecular marker in their somatic cells, but be **homozygous for that marker in the tumour cells** due to loss of the whole chromosome or deletion of that marker. This phenomenon is known as **loss of heterozygosity**, and is a common feature of mutated tumour suppressor genes.

Apoptosis

Apoptosis is **programmed cell death**. Two reasons why a cell may be forced to commit suicide are for developmental purposes and the prevention of tumour progression.

It is known that there are groups of interacting genes in a cell which are programmed to detect DNA damage or overgrowth caused by mutated oncogenes or TS genes. Instead of allowing the faulty cancer cell to divide and develop into a clone of cancer cells (and then a tumour), these genes induce the faulty cell to instigate apoptosis before the damage grows out of control.

One of these genes, *TP53*, is so frequently mutated in human cancer that its normal role has been extensively studied; this gene must be particularly important.

There is always a delicate balance between cell growth promoted by the cell cycle, and cell death, promoted by apoptosis. In some circumstances, cancer could be seen as a lack of cell death due to failure of apoptosis.

The role of p53

The **gene** ***TP53*** (*t*umour *p*rotein *53*) codes for the **protein product TP53**, which is a transcription factor. The gene locus is 17p13.1. TP53 is thought to have three roles in normal cell function; it is involved in:

- **cell cycle checkpoints and genetic stability;**
- **apoptosis;**
- **differentiation and development.**

TP53 gets its name from the molecular weight of the protein product, which is 53 000 Daltons (Da). The **gene** is often written in the commonly accepted form ***p53***; the corresponding **protein** nomenclature is then **p53**.

As a transcription factor, p53 is involved in the cell cycle by regulating genes which cause progression through that cycle. It appears to monitor the cell for DNA proof-reading or repair errors. When the genomic DNA is damaged, which may lead to tumour growth, p53 can activate genes that arrest faulty cells in G_1 and can promote their death or apoptosis.

p53 may also participate in the G_2 checkpoint, as there is evidence from mice that lack of p53 disrupts the cell's ability to monitor correct spindle assembly.

p53 probably exerts its effects on many genes, but two examples known to be targeted by p53 are *BCL2* and *BAX*. Whereas **overexpression of BAX accelerates apoptosis, BCL2** inhibits BAX, and therefore **inhibits apoptosis**. Expression of p53 causes a decrease in BCL2 and an increase in BAX. As the tumour cells are killed early on, **apoptosis** essentially **prevents tumour progression** (Figure 8.5).

BCL2 is a member of a family of cell death regulators. Its locus is 18q21, and it is expressed in cells which need a long lifespan (so that apoptosis is undesirable) such as memory B cells.

In follicular B-cell lymphoma, the translocation t(14;18)(q32;q21) creates a *BCL2*/immunoglobulin fusion gene. The immunoglobulin promoter ensures expression of BCL2 and a survival advantage in tumour cells.

The third role of p53 is to trigger cell differentiation which subsequently restricts the proliferation of genetically damaged cells. Lack of p53 has been known to cause failure of neural tube closure in mice.

The *p53* gene is usually regarded as a TS gene, as *p53* disruption causes loss of TS function: apoptosis does not occur and unregulated tumour progression ensues. Some mutations can lead to the p53 protein acting as a tumour promoter.

p53 is an unusual gene, however, in that it can also act in a dominant manner. The mutant protein product can act in a **dominant negative** fashion in the presence of normal p53 by forming complexes with and inhibiting the wild type (normal) protein.

p53 and cancer

Mutations in *p53* are the most frequent secondary changes in many cancers; point mutations and deletions of this gene occur in 70% of all tumours. *p53* mutations are found in 60% of breast tumours, and are frequently found in lung cancer and colorectal cancer.

Because p53 can stop cell proliferation by its action on the cell cycle and stop replication of damaged DNA, it has been called the **guardian of the genome**.

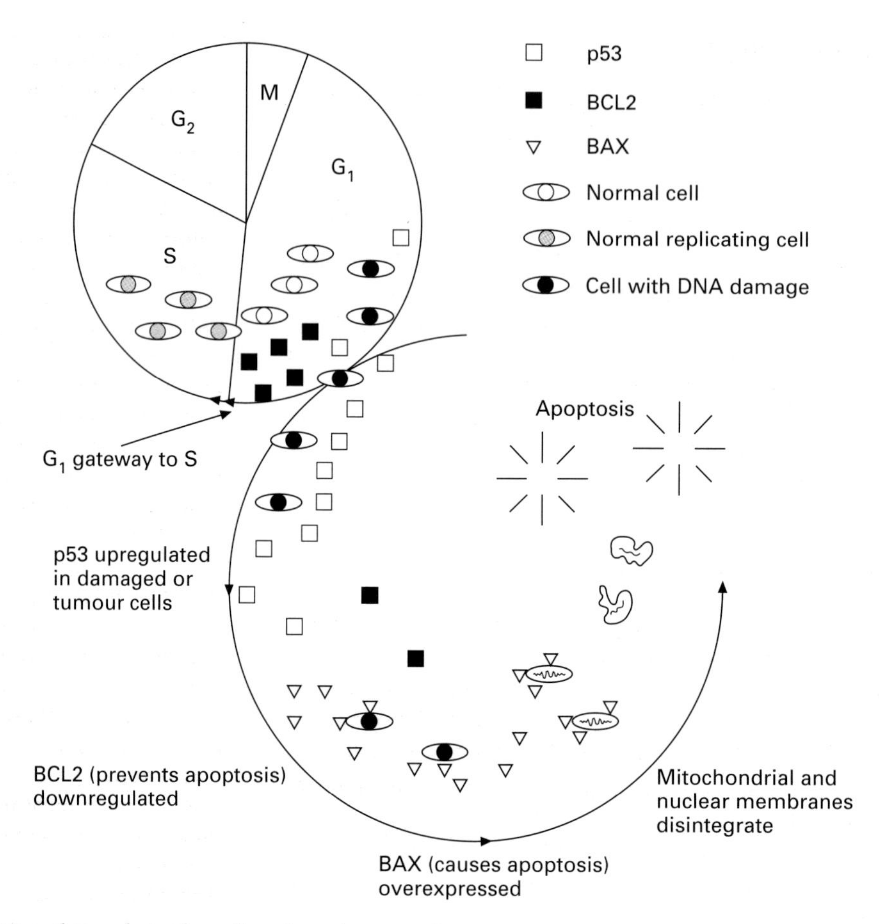

Figure 8.5 *p53 and its role in the cell cycle and apoptosis*

Li-Fraumeni syndrome is an autosomal dominant disorder whereby *p53* point mutations in one allele are inherited constitutionally (i.e. through the germline). One half of carriers develop different kinds of cancers by the time they are 30 years old (compared with 1% of the general population).

Summary

TS genes appear to act in a recessive fashion such that both alleles have to be dysfunctional or deleted before control of cell suppression is lost – resulting in tumour growth.

If an oncogene is compared to a car accelerator, then TS genes could be seen as a pair of brakes, both of which must be lost before the car (or cell!) goes out of control. TS gene mutations may be either constitutional or somatic.

Mismatch repair genes

A third class of genes associated with cancer, which were originally seen in yeast and bacteria, have now been found in humans. In the bacterium *E. coli* these are known as mutator genes and in humans mismatch repair genes. These genes are believed to code for enzymes which can 'proof-read' DNA and are therefore able to detect mismatched base pairs. The mismatching may arise through DNA replication errors or have an environmental cause.

Inactivation of mismatch repair genes leads to **instability of the genome**, thus increasing the risk of further mutations. In accordance with the guardian/gatekeeper analogy, these genes have been called **caretakers.** Using the car analogy, mismatch repair genes have been compared by Strachan and Read to a vehicle which goes out of control because some of the nuts and bolts are defective.

Chromosomes in malignancy

Probably without exception, tumour tissue contains mutations. Sometimes those mutations comprise or include changes in the karyotype. Many different types of chromosome abnormalities may be encountered, including numerical aneuploidy and polyploidy, and structural alterations such as translocations and deletions. Sometimes a tumour karyotype contains numerous abnormalities, more than could be tolerated by normal cells.

A chromosome abnormality may be a **primary** mutational event, responsible for the characteristic loss of differentiation and uncontrolled cell proliferation typical of a tumour. Alternatively it could be a **secondary** event, with poor cell cycle control facilitating errors in chromosome replication and irregularities of mitosis. It is important to understand that pathogenic chromosome abnormalities in tumours are **acquired** and not **constitutional**; in other words, the chromosome abnormality occurs only in the tumour, while normal cells in the rest of the body have a normal karyotype.

Primary chromosome abnormalities include translocations that fuse parts of two different genes to make a chimeric gene coding for a novel protein with growth regulating properties. Other translocations change the location of important growth regulating genes such that their expression is altered, and there are also examples of loss or gain of chromosomes, or parts of chromosomes, that alter the level of expression of vital genes. Overexpression of an oncogene (Figure 8.1), or deletion of a tumour suppressor gene (Figure 8.3), can lead to abnormal differentiation or cell growth.

In certain situations, chromosome analysis is an important component in identifying the exact type of tumour, often with implications regarding the prognosis for the patient and appropriate strategies for treatment. This applies particularly to **leukaemia,** malignancy involving blood cells and the blood-forming cells of the bone marrow. There are many types of mature blood cell,

The involvement of developmental genes in cancer has been suspected for a long time; the **patched** (*PTCH*) gene which we met in Chapters 2 and 3 is also a TS gene, producing a cytoplasmic protein which is important in the hedgehog signalling pathway.

Normally the hedgehog morphogen is bound by the PTCH receptor, but if *PTCH* is lost or mutated, other gene products are abnormally expressed due to the presence of the hedgehog morphogen. As PTCH is cytoplasmic, the first effects of a faulty *PTCH* may be abnormal adhesion of cells such that they heap up and proliferate. This then provides a target for other genetic faults.

It has been suggested that *PTCH* may be an example of a **gatekeeper** gene. Kinzler and Vogelstein suggest that gatekeeper genes directly regulate tumours either by inhibiting growth or promoting death. If a gatekeeper is inactivated, a genetic threshold is passed and the tumour process begins – at first in one cell of a particular tissue, closely followed by expansion from that cell.

The gatekeeper may be tissue specific – even if other proto-oncogenes are hit, as long as the gatekeeper of that tissue is intact, progression should not occur.

Because mutation or loss of *PTCH* results in basal cell carcinoma (BCC), **patched** may be the **gatekeeper of BCC** – or even of all common skin cancers.

Another example of a gatekeeper is found in the gene for Von Hippel–Lindau (VHL) disease, where mutations result in cancer of the kidneys.

In the autosomal dominant disease **h**ereditary **n**on-**p**olyposis **c**olon **c**ancer (HNPCC), **new alleles** may be seen in the **tumour tissue** when polymorphic microsatellite markers are used. This is called **microsatellite instability**.

There are six human mismatch repair genes. Two are due to gene mutations in the same chromosome regions as those found in HNPCC patients:

- *hMSH2* at 2p15-22
- *hMLH1* at 3p21.3

In some families individuals develop colon cancer at a very young age. It may be that this predisposition is constitutionally inherited and may involve mutations in mismatch repair genes.

Other examples of caretaker genes include the breast cancer genes *BRCA1* and *BRCA2*, and possibly the *ATM* gene of ataxia telangiectasia.

Breast cancer is believed to be due to mutations at two major gene loci:

- The *BRCA1* locus is 17q21 and is associated with breast and ovarian cancer.
- The *BRCA2* locus is 13q12-13 and is associated with female and male breast cancer.

However, although these TS mutations are inherited in *BRCA1* and *BRCA2*, they are absent in sporadic tumours. This implies that they do not follow Knudson's hypothesis and is consistent with the theory that breast cancer genes are caretaker genes whose normal gene products (which are

each with its specific function, derived by sequential differentiation from unspecialized stem cells. A specific mutation, often a chromosomal change, in a stem cell causes leukaemia, the exact type of leukaemia depending on the stage of differentiation disrupted by the particular genetic change.

Examples of chromosome abnormalities valuable in diagnosis include a 9;22 translocation, called the **Philadelphia translocation,** characteristic of **chronic myeloid leukaemia** (CML). At the molecular level this translocation involves fusion of parts of two genes, *ABL* and *BCR*, to produce a new active protein with particular growth regulating properties.

Acute myeloid leukaemia (AML) is often classified into seven subtypes, M1 to M7, depending on the morphology of the leukaemic cells; specific chromosome abnormalities are associated with particular subtypes. Taken in conjunction with the abnormal cell morphology, the chromosome abnormality is a dependable indicator of the correct diagnosis. An 8;21 translocation is found in subtype M2, a 15;17 translocation in M3, and a pericentric inversion of chromosome 16 in M4. These three chromosome changes tend to be associated with a relatively favourable prognosis. Rapid confirmation of the 15;17 translocation is important, as patients with this rearrangement and AML M3 are susceptible to coagulation disorders and require a special regimen of treatment.

There are other chromosome abnormalities in acute myeloid leukaemia not associated with specific subtypes. Rearrangements of the MLL locus on chromosome 11q23, deletions of the long arm of chromosome 5 and deletions or loss of chromosome 7 are also found. With these abnormalities, many of the patients have had previous exposure to cytotoxic agents, for example treatment for some other malignant condition, and often the prognosis is relatively poor.

Acute lymphoblastic leukaemia (ALL) is more frequent in children, and also has its own characteristic chromosome changes providing valuable information with respect to diagnosis and prognosis. Amongst these, **hyperdiploidy** with more than 50 chromosomes is frequent: the chromosomes gained are not always exactly the same but most often include 4, 6, 10, 14, 18 and 21. The Philadelphia translocation is sometimes seen in ALL, conferring a poor prognosis. A 4;11 translocation is found primarily in infants, and involves the MLL locus, mentioned above with regard to AML. MLL, incidentally, stands for Mixed Lineage Leukaemia. Only recently another translocation has been discovered, involving chromosomes 12 and 21, and it is now recognized to be extremely common in ALL. As this rearrangement involves an exchange of small, pale-staining segments of similar length from the tips of the short arm of 12 and the long arm of 21, it is impossible to see using ordinary banding, and is only visible by FISH (see Chapter 6).

As far as solid tumours are concerned, cytogenetic analysis is less often useful in diagnosis. Tumour tissue can be difficult to grow in culture, and karyotypes often contain complex abnormalities making it difficult to identify specific relevant rearrangements. However, chromosome analysis can be valuable in some childhood tumours such as **Wilms tumour, neuroblastoma, Ewing sarcoma** and **rhabdomyosarcoma,** where specific diagnostic changes may be found.

When a chromosome study is undertaken on malignant tissue, the starting point is usually conventional G-bonded analysis. There are problems, insofar as the tissue often yields poor quality chromosomes, few cells that are analysable, and very complicated chromosome changes. Increasingly, FISH is being used to supplement the analysis. For example, FISH probes for the *BCR* and *ABL* loci can confirm the Philadelphia translocation not only in metaphase chromosomes but also in interphase nuclei, where *BCR* gives a green signal, *ABL* a red signal, and the translocation a red/green fusion signal (see Figure 6.4). A similar strategy is available for confirmation of the 12;21 (ALL) translocation. A FISH probe spanning the MLL locus is split when there is a translocation, resulting in one large signal on the normal chromosome 11, and two smaller signals, one on each of the translocation products.

expressed at G_1/S) are essential co-factors in DNA repair.

The risk to a British woman of developing breast cancer is between 1/8 and 1/12. This risk increases if there are close female relatives with breast cancer. Around 5–10% of breast cancer is familial.

If a familial mutation has been inherited by an individual, counselling is advisable, as the woman in question has an 80–90% risk of the cancer developing (not 100%). If she does not inherit the mutation, her overall risk only falls back to that of the general population – she is not completely free of risk.

The mutations in breast cancer are screened for using DNA analysis (see Chapter 7). Strict criteria must be met before an individual is accepted onto a screening programme; this usually requires the female consultand to have two (or sometimes three) affected first degree relatives.

The multistep nature of cancer

The development of cancer can be seen as a series of steps, each of which gives a growth advantage to a cell. One particular colon cancer has been used as a model as it has a well-defined pattern of progression. Familial adenomatous polyposis coli (**FAP** or FAPC) is an autosomal dominant colon cancer, which first appears in an affected person in late childhood to teenage years. Hundreds or thousands of growths called adenomatous **polyps** develop in the large intestine. In this disorder it is almost certain that one or more polyps will become malignant, passing through the early, intermediate and late stages of **adenoma,** finally culminating in malignant **carcinoma.**

Fearon and Vogelstein (1990) described each progressive stage with respect to the interaction of the various oncogenes and TS genes believed to be implicated in each step (Figure 8.6). The gene which is mutated or lost in FAP is called *APC* (adenomatous polyposis coli), located at 5q21. This may be lost constitutionally, and although only involving a single copy, it is sufficient to initiate polyp formation by giving the colon epithelial cells a proliferative advantage. The **APC** gene acts like a TS gene in that the final stage carcinomas often show LOH of *APC*, although this is rarely seen in early stages. Thus the distinction between dominant oncogenes and recessive TS genes (as in *p53*) is not as distinct as previously thought.

The steps leading to colon cancer may therefore begin with the loss or mutation of *APC*. Each stage in adenoma progression is then

Conventional DNA polymerases cannot replicate telomeric sequences, which cap chromosome ends. Each time somatic fibroblast cells divide, telomeres shorten and the cells gradually enter senescence. An enzyme called **telomerase** is present in germ cells, which is able to add telomere repeats on to chromosome ends. Although its activity is not detected in most normal somatic cells, it is **detectable in many immortal cell lines and human tumours**.

There is evidence that in some epithelial cell types telomerase can be activated by oncogenes such as *MYC*, and by increasing telomere length these oncogenes extend the lifespan of the cells,

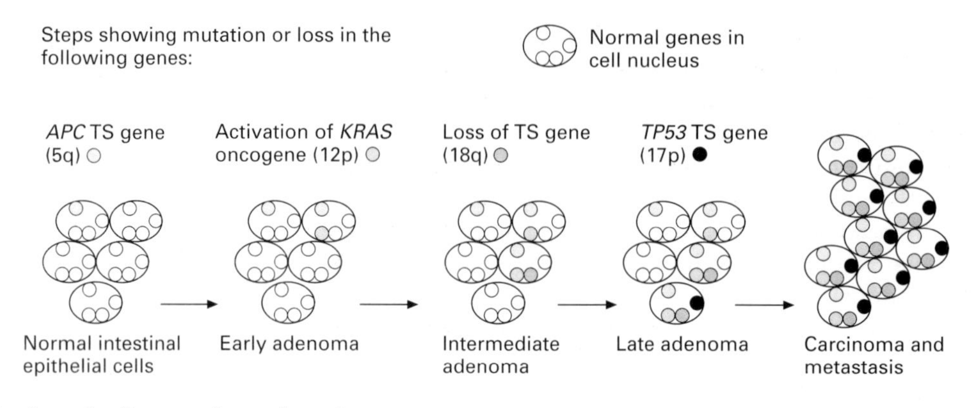

Figure 8.6 *Steps leading to colorectal carcinoma*

leading to immortalization and tumour formation.

It has been suggested that telomerase might be employed as an 'anti-ageing' enzyme, or that an 'anti-telomere' drug could be synthesized to fight cancer, despite evidence that other factors may also be required for cell immortality *in vivo*.

Looking at the gene mutations found at each stage of adenoma/carcinoma can support the multistep theory.

- It is common to see *KRAS* mutations in intermediate to late adenomas and carcinomas (50%), but rare in the early ones (10%).
- It is common to see LOH of 18q in late adenomas and carcinomas (50%), but rare in the early stages.
- Many carcinomas have *p53* mutations, which are not seen in adenomas.

dependent on another gene fault such as activation of the *KRAS* oncogene, loss of a TS gene on 18q (possibly *SMAD4*), and loss or mutation of *p53*, which allows progression from G_1 to S phase instead of apoptosis. Any mutations in the mismatch repair genes act in the background, predisposing the adenomas to progress by increasing the general mutation rate.

Summary

Although it would be ideal to have a universal theory for the development of cancer, it is difficult to find a general explanation for neoplasias, which range from the inherited purely genetic cancers such as FAP to the acquired cancers such as leukaemia.

However, the human body does seem to have multiple safeguards against DNA damage leading to carcinogenic changes. It may be that the caretaker genes such as the mismatch repair genes require two hits to make the cell genome unstable, followed by another two hits of a gatekeeper gene to begin the process of tumour formation in a particular cell type. Even then, we have the 'guardian angel' gene *p53*, which would also need to be knocked out in order for tumour cell proliferation to proceed.

It is now believed that models such as that for colon cancer, based on a linear progression, may be an oversimplification of the many different routes and changes seen in various cancers. The neoplastic process may change pathways or even regress before proceeding to its final conclusion, requiring a more flexible theory of the genetic changes that lead to cancer.

Suggested further reading

Davies, K. (1996). Cancer and development patched together. *Nature Genetics*, **13**, 258.

Fearon, E.R. and Voglestein, B. (1990). A genetic model for tumorigenesis. *Cell*, **61**, 759–767.

Gottlieb, T.M. and Oren, M. (1996). p53 in growth control and neoplasia. *Biochimica et Biophysica Acta*, **1287**, 77–102.
Hesketh, R. (1997). *The Oncogene and Tumour Suppressor Gene Facts Book*, 2nd Edn. Academic Press.
Kaelin, W.G. and Maher, E.R. (1998). The *VHL* tumour-suppressor gene paradigm. *Trends in Genetics*, **14**(10), 423–426.
Knudson, C.M. and Korsmeyer, S.J. (1997). *Bcl-2* and *Bax* function independently to regulate cell death. *Nature Genetics*, **16**, 358–363.
Lane, D.P. (1992). p53, guardian of the genome. *Nature*, **358**, 15–16.
Lane, D.P. (1998). Awakening angels. *Nature*, **394**, 616–617.
Macdonald, F. and Ford, C.H.J. (Eds) (1996). *Molecular Biology of Cancer*. Coronet Books.
Mueller, R.F. and Young, I.D. (1998). The genetics of cancer, in *Emery's Elements of Medical Genetics*, 10th Edn. Churchill Livingstone.
Sidransky, D. (1996). Is human *patched* the gatekeeper of common skin cancers? *Nature Genetics*, **14**, 7–8.
Strachan, T. and Read, A.P. (1999). *Human Molecular Genetics 2*, Ch. 18. Bios Scientific Publishers Ltd.

Self-assessment questions

1. What is the name given to tumours that are acquired during one's lifetime rather than inherited? Name three environmental factors which might contribute to the development of malignancies.
2. Why are growth factor genes likely to show oncogenic properties when mutations occur?
3. Why is the t(8;14) translocation in Burkitt lymphoma significant? (Hint: What happens to one of the translocated genes when it changes position?)
4. There is a syndrome called the WAGR complex (Wilms' tumour, aniridia, gonadoblastoma and mental retardation). A baby is born with aniridia (lack of the iris, the coloured part of the eye). The clinical geneticist asks for chromosome 11 to be checked – why? The baby's kidneys are going to be scanned regularly for the first 2 years. Why is the baby being checked at such an early age? What sort of cancer gene may be involved?
5. A protein called MDM2 regulates p53 by degrading it if there is too much. What would happen to the amount of p53 if the action of MDM2 was blocked by DNA damage? What effect would this have on the affected cells?
6. Why do you think that the *RB1* and *APC* genes are always said to act in a dominant manner in familial cases, when they are supposed to be 'recessive' TS genes?
7. Why have mismatch repair genes been described as caretakers? Briefly explain what might happen if they are mutated.

Key Concepts and Facts

What is Cancer?

- Cancer is a loss of cell cycle control such that there is inappropriate or increased cell proliferation. There is a variable genetic component which may require additional environmental factors for the development of full malignancy.

Oncogenes

- Control is lost over genes responsible for cell growth or proliferation. Oncogenes behave in a dominant manner and often arise from point mutations, translocations or inversions, and represent a gain of function.

Tumour Suppressor Genes

- Loss of suppression of cell growth occurs when both alleles of a TS gene are lost. Knudson's two hit hypothesis characterizes TS genes such that the first hit is either constitutional or somatic, while the second is somatic. The second hit is predicted to be found only in the tumour, as shown by LOH studies. TS genes act in a recessive manner, and are often due to deletions or point mutations, and represent a loss of function.

Apoptosis

- Programmed cell death is controlled by many genes; one of the best characterized is *p53*. Failure of *p53* in one of its roles (guarding against abnormal replication of damaged cells which would normally self-destruct) may result in cancer.

Mismatch Repair Genes

- Mismatch repair genes proof-read replicated DNA for errors; a sign of mutation in one of these genes is microsatellite instability, where new alleles appear that should have been corrected. Abnormal mismatch repair genes increase the risk of other genes mutating, in turn increasing the risk of cancer.

Cancer and Acquired Chromosome Abnormalities

- DNA mutations may be inherited constitutively or acquired sporadically. In the examples shown by some leukaemias and solid tumours, the specific acquired chromosome abnormalities are consistent and predictive.

Theories

- Although there are examples of cancer occurring in a stepwise linear progression, it is possible that malignancy is the final result of many disparate pathways, requiring the mutation of caretaker and gatekeeper genes, together with that of the cell cycle checkpoints such as the guardian of the genome *p53*.

Part Three:

Prevention of Disease

Chapter 9
Prenatal diagnosis

Learning objectives

After studying this chapter you should confidently be able to:

Explain what is meant by prenatal diagnosis.

Discuss the criteria for effective prenatal diagnosis.

Discuss how pregnancies at risk of foetal abnormality are identified.

Describe the methods of prenatal diagnosis.

Outline the problems and limitations associated with each method.

Describe the principles of screening programmes.

Introduction

The desire of all parents is that any children they have shall be healthy and normal. However, some children are born with varying degrees of problems. About 14% of infants are born with a single minor **malformation** such as an ear tag, 3% have a single major malformation such as isolated spina bifida and 0.7% have multiple malformations. Some children have mental handicap with learning difficulties, behavioural problems or autism.

A malformation is a primary error of normal development or morphogenesis of an organ or tissue. All malformations are congenital, i.e. present at birth, although they may not all be diagnosed until later in life if microscopic or if internal organs are involved.

The causes of many malformations are unknown, but genetics plays a part in over one-third where a cause is known. This may be due to a **chromosome** abnormality or to a **Mendelian disorder.** Mendelian disorders are those that are a result of a single mutant gene and follow the simple patterns of autosomal or sex-linked and recessive or dominant inheritance (see Chapter 4). Over 5000 Mendelian disorders are known in humans.

The **aetiology** of major congenital malformations is as follows:

Idiopathic (no known cause)	60%
Multifactorial	20%
Monogenic	7.5%
Chromosomal	6.0%
Maternal illness	3.0%
Congenital infection	2.0%
Drugs, X-rays, alcohol, teratogens	1.5%

Prenatal diagnosis is the detection or exclusion of abnormality in the foetus during pregnancy. It is important in detecting and preventing genetic disease. It is not about eugenics and creating perfect babies but about giving parents information and allowing them the choice as to whether or not they wish to continue with a pregnancy in which severe abnormality has been detected. Most parents would terminate an affected pregnancy but some choose to

A normal pregnancy lasts for 40 weeks calculated from the first day of the last menstrual period (LMP). Pregnancy is divided into three parts of approximately 3 months each, called the **first**, **second** and **third trimesters**.

Phenylketonuria is an autosomal recessive disorder with deficiency of the enzyme phenylalanine hydroxylase, which normally converts the amino acid phenylalanine to tyrosine. If not treated the toxic accumulation of phenylalanine will result in a child with severe mental retardation. Removal of phenylalanine from the diet provides effective treatment. PKU is an example of an **inborn error of metabolism**, a genetically inherited metabolic defect, which results in deficient enzyme production or synthesis of an abnormal enzyme.

The criteria for prenatal diagnosis can be summarized as:

- High risk of abnormality.
- Severe disorder.
- No treatment available.
- Early treatment advantageous.
- Reliable test.

use the information to prepare for the birth of an affected child. When the foetus is found to be normal it allows the parents to continue the pregnancy reassured.

Criteria for prenatal diagnosis

Before undertaking prenatal diagnosis several factors need to be considered.

- The **pregnancy** has been **identified** as being at **high risk of foetal abnormality**.
- The **disorder must be severe**. Many genetic disorders lead to death *in utero*, or in infancy or childhood. However, many can be compatible with survival for years with severe handicaps such as Down syndrome, spina bifida and cystic fibrosis.
- There is **no available treatment**. Many genetic disorders cannot be treated even if the basis for the disease is known. Research into the use of gene therapy may help to change this.
- The disorder can be **treated effectively**, as in phenylketonuria (PKU), or if early treatment such as surgery is beneficial. Prenatal diagnosis will allow the child to be delivered in a centre with specialist services if appropriate and early treatment commenced.
- The **prenatal test** must be **reliable** enough for a decision to be made. Over 200 Mendelian disorders can now be detected prenatally by DNA or biochemical analysis and the number is rapidly increasing.

Identification of pregnancies at risk

There are various ways of identifying pregnancies at risk of foetal abnormality. These include:

- Advanced maternal age (usually over 35 years).
- Carriers of autosomal recessive disorders identified by population screening programmes.
- Maternal serum biochemistry.
- Ultrasonography.
- Previous child affected with a Mendelian disorder.
- Family history of a Mendelian disorder.
- Previous child with a chromosome abnormality.
- Inherited chromosome abnormality.

Maternal age

In 1934 Penrose first recognized that children with Down syndrome are more often born to older parents. This was later attributed to the age of the mother, with paternal age having little effect. This is thought to be due to the effect on the oocytes, which arrest in meiosis I from before birth until released at ovulation. Suspension of meiosis at this stage makes the oocytes susceptible to damage from the environment such as drugs and radiation.

Trisomy arises by meiotic non-disjunction, most commonly in the first meiotic division in the mother (see Chapter 5). Non-disjunction in female meiosis II and in male meiosis are much less common mechanisms of non-disjunction.

The development of the techniques of amniocentesis and tissue culture to provide chromosome preparations for foetal karyotyping enabled prenatal diagnosis to be offered to detect pregnancies with Down syndrome from the 1960s onwards.

Trisomy 21, Down syndrome, is the commonest chromosome abnormality with an incidence of 1 in 650 at birth. A child with Down syndrome can be born to parents of any age; however, the risk rises steeply once the mother's age is over 35 years (Figure 9.1). The risks in the first and second trimester are higher since many affected pregnancies will be lost as miscarriages and stillbirths. Approximately 75% of all Down syndrome conceptions are lost prior to term.

Prenatal diagnosis based on maternal age has only limited success in detecting pregnancies with Down syndrome because, although the risk of Down syndrome is higher in older mothers, most babies are born to mothers less than 35 years of age. The detection rate obtained using maternal age alone is only in the region of 30%.

The figures for approximate risk of trisomy 21 at birth according to maternal age are as follows:

All ages	1 in 650
Age 20	1 in 1540
Age 30	1 in 890
Age 40	1 in 100
Age 44	1 in 40

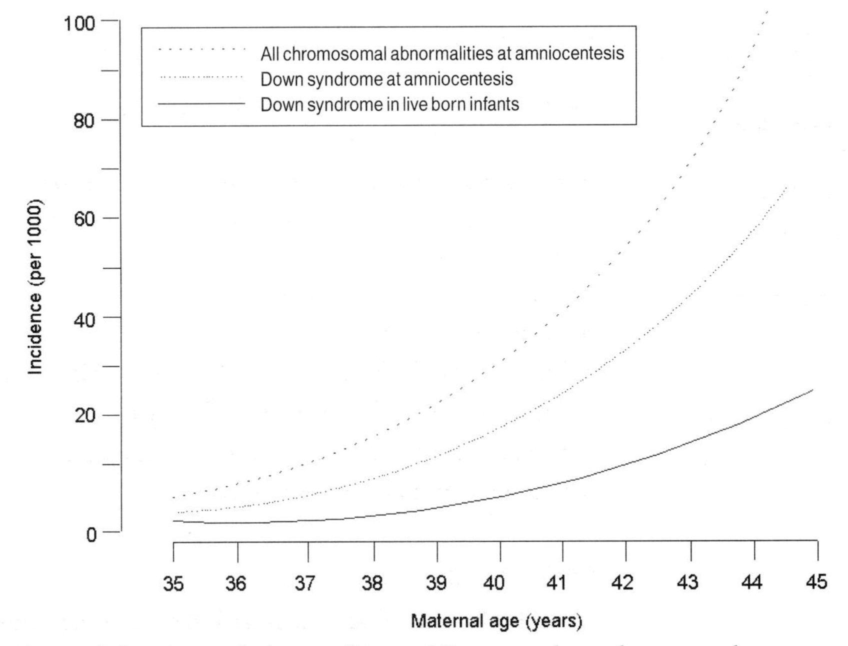

Figure 9.1 *Incidence of chromosomal abnormalities and Down syndrome by maternal age*

The **principles** underlying a **screening programme** can be summarized as follows:

- Method available.
- Clearly defined disorder.
- Appreciable frequency.
- Advantage of early diagnosis in terms of allowing treatment or prenatal diagnosis.
- Low false **positive rate (specificity)**.
- Low false negative **rate (sensitivity)**.
- Benefits outweigh the costs.

- **Specificity** is the extent to which a test detects only affected individuals. False positives are unaffected individuals detected.
- **Sensitivity** is the proportion of cases detected. False negatives are the affected cases that are missed.
- **Detection rate** is the number of abnormals detected.

Some examples of **autosomal recessive disorders** for which **population screening** is suitable are given below.

- α-Thalassaemia: China, South-Eastern Asia and Mediterranean countries.
- β-Thalassaemia: Indian sub-continent and Mediterranean countries.
- Sickle cell disease: Afro-Caribbeans.
- Tay–Sachs disease: Ashkenazi Jews.
- Cystic fibrosis: Western European Caucasians.

The increased risk associated with maternal age is also observed for other autosomal trisomies (of which only trisomy 13 and 18 are seen at birth), for marker chromosomes and for additional copies of the X chromosome, but these are much less frequent findings at birth. Types of chromosome abnormalities are described in more detail in Chapter 5.

Population screening

Population screening entails the testing of a whole population in order to detect those at risk of a genetic disease in themselves or their offspring. This, combined with genetic counselling, allows couples to be forewarned of any risk that each of their children will be affected. However, this approach is not appropriate for all genetic diseases, as certain principles need to be observed.

Although many genetic diseases are well defined, they are too rare to merit a whole population screening programme. Exceptions to this occur due to the relative high frequency of certain disorders in particular ethnic groups. The introduction of certain screening programmes has been 'successful' in that they have led in some areas to a reduction in the birth of affected children following their detection by prenatal diagnosis and termination of the pregnancy.

Screening using maternal serum biochemistry

During pregnancy the foetal and maternal circulations do not mix but exchange of some chemicals can occur, the presence of which can give an indication of foetal well-being. One of these chemicals is alpha fetoprotein (AFP). Maternal serum AFP (MSAFP) concentration is used to screen for neural tube defects such as spina bifida and for Down syndrome.

Maternal serum biochemistry used to screen for neural tube defects

As long ago as 1972, it was reported that a mother carrying a baby with a neural tube defect (NTD) such as spina bifida had an increased level of AFP in her serum. This is usually 2.5 standard deviations above the mean value expected for the gestation. The increase is due to leakage of foetal serum into the amniotic fluid from exposed foetal capillaries. This observation led to the introduction of a highly successful screening programme for the detection of NTDs. Over 80% of open spina bifida cases and almost 100% of anencephalic foetuses can be detected.

There is some overlap between the values of AFP found in normal pregnancies and those with a foetus with a neural tube defect (Figure 9.2). The presence of a neural tube defect is not the only cause for a raised value for AFP in maternal serum. If a high value is

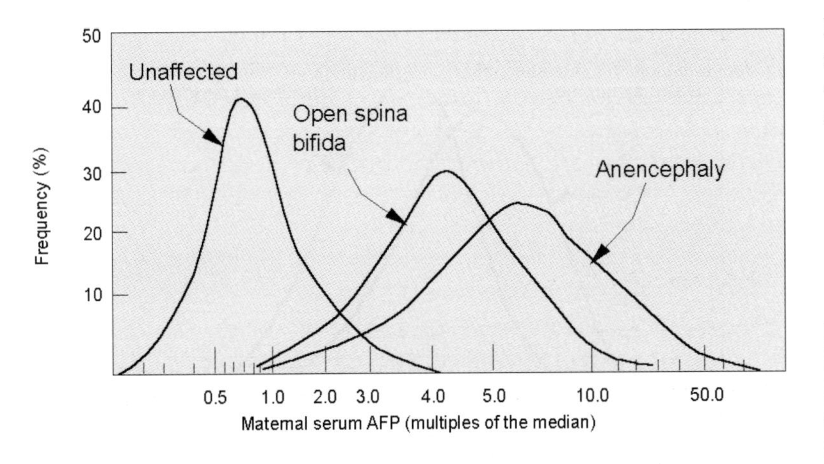

Figure 9.2 *Maternal serum AFP in normal pregnancies and pregnancies affected with open spina bifida and anencephaly*

obtained a repeat sample may be tested and an ultrasound scan of the foetus undertaken to identify the reason for the result.

The cause of NTDs is not known but it has been observed that a diet high in folic acid taken from preconception reduces the risk of a child developing a NTD, and this has led to a significant reduction in their incidence. This is a major breakthrough in that it reduces the occurrence of an abnormality rather than having to consider termination once the abnormality has been detected during pregnancy.

AFP is the major foetal plasma protein, synthesized first in the foetal yolk sac and later in the liver. It is structurally but not antigenically similar to albumin. Some AFP is excreted via the foetal urine into the amniotic fluid from where some diffuses into the maternal circulation. The levels vary during gestation, with a peak in foetal plasma at 12–14 weeks and in maternal serum at 30 weeks. The levels in maternal serum are 1000 times lower than in the amniotic fluid.

Maternal serum biochemistry used to screen for Down syndrome

The observation that **low** levels of **MSAFP** were associated with **increased risks** of Down syndrome was first made in the early 1980s. As with the values associated with neural tube defects, there is an overlap between the values found in normal pregnancies and those where the foetus has Down syndrome (Figure 9.3). The combination of this assay with the measurement of human chorionic gonadotrophin and sometimes oestriol has led to an alternative screening strategy for Down syndrome.

Human chorionic gonadotrophin (HCG) is secreted by the developing trophoblast and later by the chorion and placenta. It can be detected in maternal serum from 3–4 weeks post LMP. The levels increase in the serum exponentially during the first trimester, reaching a plateau in the second and third trimesters. Levels of HCG in maternal serum have been found to be elevated up to twice the normal expected values in the second trimester in pregnancies affected with Down syndrome.

Oestriol is a steroid hormone produced by the syncytiotrophoblast from foetal precursors. It is assayed in the unconjugated form. The values found in the maternal serum, for AFP and HCG (double test) or with oestriol also (triple test), are combined with maternal

A NTD is the defective closure of the neural tube during early development. The neural tube appears at 20 days and is mostly closed by 23 days. Failure of closure at the cephalic (head) end produces anencephaly and lower down produces spina bifida.

The incidence at birth shows marked geographic variation, from 1 in 1000 births in the USA to 8.6 in 1000 births in Eire.

Possible reasons for raised levels of MSAFP are as follows:

- Neural tube defects.
- Incorrect gestational age.
- Intrauterine foetal death.
- Multiple pregnancy.
- Threatened miscarriage.
- Abdominal wall defect.
- Congenital nephrotic syndrome.

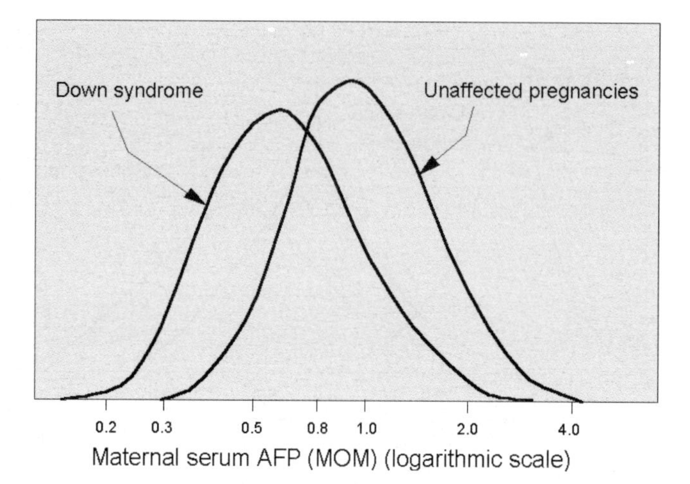

Figure 9.3 *Maternal serum AFP expressed as multiples of the median (MOMs) in normal pregnancies and pregnancies affected with Down syndrome*

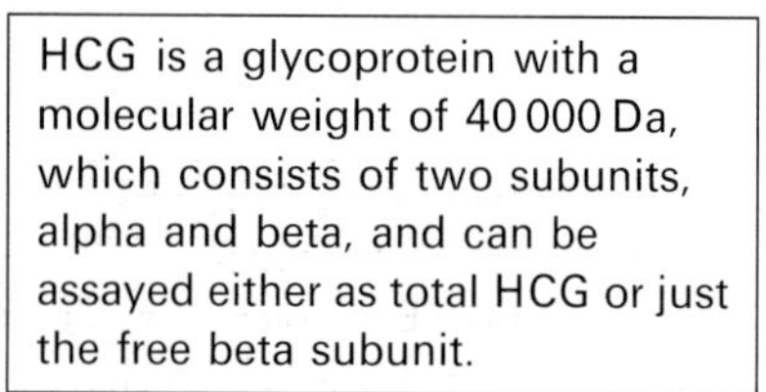
HCG is a glycoprotein with a molecular weight of 40 000 Da, which consists of two subunits, alpha and beta, and can be assayed either as total HCG or just the free beta subunit.

Since the levels of AFP, HCG and oestriol all vary with gestation the values are expressed as **multiples of the median (MOMs)**. A MOM of 1 means that the values are as expected for the gestation, a value of 2 MOM is raised by twice the expected level, and a value of 0.7 is 70% of the expected level.

MOMs in pregnancies affected with Down syndrome are as follows:

HCG	2.05
AFP	0.75
Oestriol	0.73

A risk of 1 in 300 means that for every 300 births, 299 would be expected to be normal and one to have Down syndrome. A ratio of 1 in 200 represents a greater risk than a ratio of 1 in 500.

age, using a complex formula on a computer programme, to give each mother an individual risk level.

This is based on the age-related risk (prior odds) and the likelihood ratio. The values of the cut-off determine the detection rate and the false positive rate. The lower the cut-off, the lower the false positive rate but the lower the detection rates also. The most widely used cut-off is a risk of 1 in 250 or 1 in 300. This is similar to the risk of a liveborn child with Down syndrome based on a maternal age of 35 years.

This cut-off gives a detection rate of about 60% (compared to 30% from maternal age alone) with a false positive rate of 5%. This is only a risk and not a definitive diagnosis. Women with a result higher than a predetermined cut-off will be offered foetal karyotyping as described later.

Ultrasonography

Ultrasonography involves using high frequency sound waves inaudible to the human ear to produce a picture of the baby. The waves are directed through a transducer into the body and are reflected back as echoes. These are fed back into a computer that uses the information to build up a series of black and white pictures which are linked to produce a moving image (Figure 9.4).

General uses of ultrasonography include checking foetal viability, diagnosis of multiple pregnancies, estimating gestation from foetal size and monitoring foetal growth during pregnancy.

Second trimester foetal anomaly scans

An ultrasound scan is usually offered to all pregnant women at between 18 and 20 weeks to screen for foetal anomalies. At this stage it is possible to detect abnormalities affecting the central nervous system, limbs, heart, kidney and gastrointestinal tract. Not

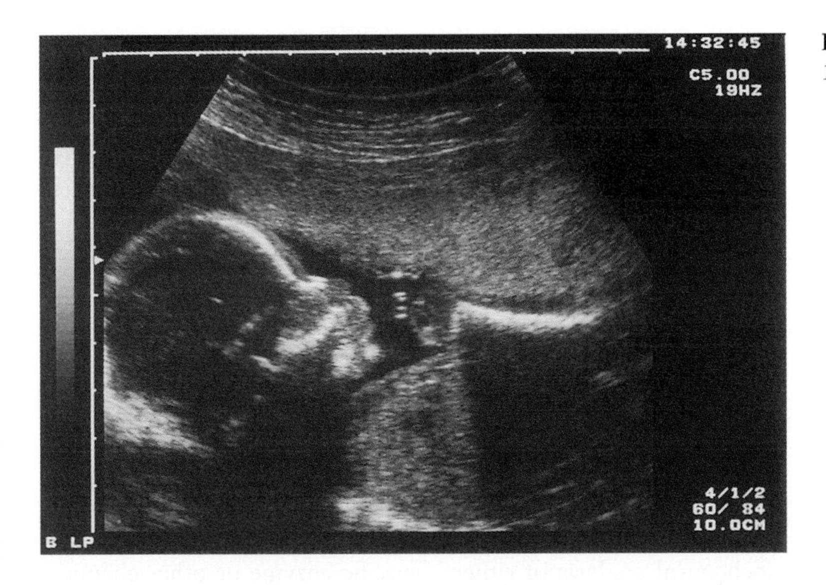

Figure 9.4 *Normal ultrasound scan at 18 weeks gestation*

all abnormalities will be detectable. Some that are detected may be only very minor and not require any action. Others may indicate the presence of an underlying chromosome abnormality, which can be investigated by amniocentesis, chorionic villus or foetal blood sampling (see later), and can allow for preparation for appropriate care after birth.

Certain **ultrasound findings** suggest a **chromosome abnormality**. Examples of such findings are given below, along with the chromosome abnormality suggested.

- Cardiac defect: trisomy 13, 18, 21 or deletion of chromosome 22.
- Choroid plexus cysts: trisomy 18, 21.
- Cystic hygroma: trisomy 13, 18, 21 or Turner syndrome.
- Exomphalos: trisomy 13, 18.
- Rocker-bottom feet, overlapping fingers: trisomy 18.
- Growth retardation: triploidy.

First trimester scan

At 11 weeks gestation the foetus is approximately 7 cm in size and the foetal heart can be seen beating. Even this early some abnormalities can be seen. One example is an abnormal collection of fluid behind the foetal neck (nuchal translucency), which can be seen between 10 and 14 weeks and which has been found to be associated with an increased risk of chromosome abnormality. The risk is related both to the thickness of the nuchal translucency and the maternal age. Measurement of the nuchal area is now being increasingly used as a method of screening for Down syndrome but requires the use of highly trained staff. The advantage is that it can be done earlier than the second trimester maternal serum biochemistry and is also reported to have a higher detection rate of up to 90%. This earlier test helps to reduce parental anxiety and allows termination to be performed much earlier in pregnancy if appropriate.

The finding of a normal karyotype in foetuses with a raised nuchal translucency thickness does not exclude other problems. There is an increasing awareness that these foetuses are at increased risk of other problems including an increased risk of intrauterine death.

The **increased risk** of **trisomy** 13, 18 and 21 from **nuchal scans** (given as size in mm) is as follows:

3 mm	× 4
4 mm	× 21
5 mm	× 26
>6 mm	× 41

Previous child with a Mendelian disorder

Unfortunately, for most couples the finding that they are carriers of an inherited disorder is not made until the birth of an affected child. The risk in subsequent pregnancies for an autosomal recessive disorder is then 1 in 4. Carriers of autosomal dominant disorders have a risk of 1 in 2. Carriers of X-linked recessive disorders have a 1 in 2 chance that a male foetus will be affected. Calculation of risks is described in Chapter 4.

At present, tests are only available for a limited number of disorders and therefore prenatal diagnosis may not always be feasible. Fortunately, this number is increasing rapidly due to advances in technology, particularly in the field of **molecular genetics.**

In order for DNA analysis to be possible it is necessary for the mutation to have been identified within a family or for closely linked informative markers to have been identified (Chapter 7). For biochemical analysis of either a specific enzyme or other chemical there must be a clear distinction between carriers and affected cases, otherwise it will not be possible to give accurate risks for the foetus (see Chapters 4 and 7). Examples of prenatal diagnosis available at present are listed in Table 9.1.

Family history of Mendelian disorders

Genetic counselling for families where a Mendelian disorder is known to exist is necessary in order to identify the risk in a pregnancy. This will be based on the mode of inheritance, population frequency of the disorder and tests to determine the carrier status of the couple. Calculation of risks is discussed more fully in Chapter 4. Genetic counselling is discussed in Chapter 10.

Table 9.1 *Examples of prenatal diagnosis available at present*

Disease	*Inheritance*	*Method of analysis*
Cystic fibrosis	AR	DNA
Hurler syndrome	AR	Biochemical
Spinomuscular atrophy	AR	DNA
Myotonic dystrophy	AD	DNA
Huntington disease	AD	DNA
Achondroplasia	AD	DNA
Haemophilia A	XL	DNA/biochemical
Fragile X syndrome	XL	DNA
Duchenne muscular dystrophy	XL	DNA
Tay–Sachs disease	AR	Biochemical
Ornithine carbamoyltransferase	AR	Biochemical
Sanfillipo syndrome	AR	Biochemical
Lesch–Nyhan syndrome	XL	Biochemical

AR = autosomal recessive; AD = autosomal dominant; XL = X-linked.

Previous child with a chromosome disorder

Parents who have had a child with trisomy 21 or any other trisomy are at increased risk of having a trisomy in subsequent pregnancies. This is about 1% above the maternal age-related risk and many of these couples request prenatal diagnosis. This increased risk may be the result of one parent being a low-level mosaic for the chromosome or there may be a genetic predisposition to non-disjunction in one parent.

Risks from any other previous *de novo* chromosome abnormality is extremely low but parents may request prenatal diagnosis for reassurance.

Inherited chromosome rearrangements

Carriers of a chromosome rearrangement are at increased risk of producing **chromosomally unbalanced** offspring, as described in Chapter 5. The risk in any individual family depends on several factors. These include the chromosomes involved and the size of the potential imbalance. Many imbalances will result in non-viable conceptions. Observation of chromosomal imbalance as an abnormal liveborn infant indicates that this is a possible outcome in any pregnancy although some unbalanced offspring may also miscarry. Some couples may choose to defer prenatal diagnosis until the second trimester since unbalanced offspring may be lost naturally early in pregnancy.

Methods of prenatal diagnosis

There are several methods available for prenatal diagnosis (Figure 9.5). These involve the removal of foetal tissue for analysis. These invasive tests carry a risk of losing the pregnancy from spontaneous miscarriage related to the procedure. The methods currently available include:

- Amniocentesis
- Chorionic villus sampling
- Foetal blood sampling.

Amniocentesis

Amniocentesis involves the removal of amniotic fluid (AF) by the insertion of a needle under ultrasound guidance through the mother's abdomen into the amniotic cavity and is usually undertaken at 14–16 weeks gestation, when 10–20 ml of fluid is removed. It is the most commonly used method of obtaining cells for foetal karyotyping, with about 40 000 procedures being performed in the UK each year, and can also be used to measure levels of alpha fetoprotein to detect NTDs and other chemicals for

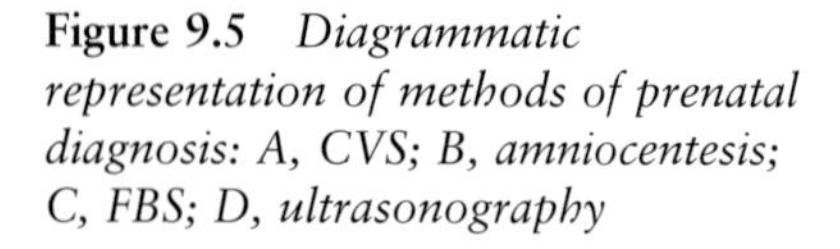

Figure 9.5 *Diagrammatic representation of methods of prenatal diagnosis: A, CVS; B, amniocentesis; C, FBS; D, ultrasonography*

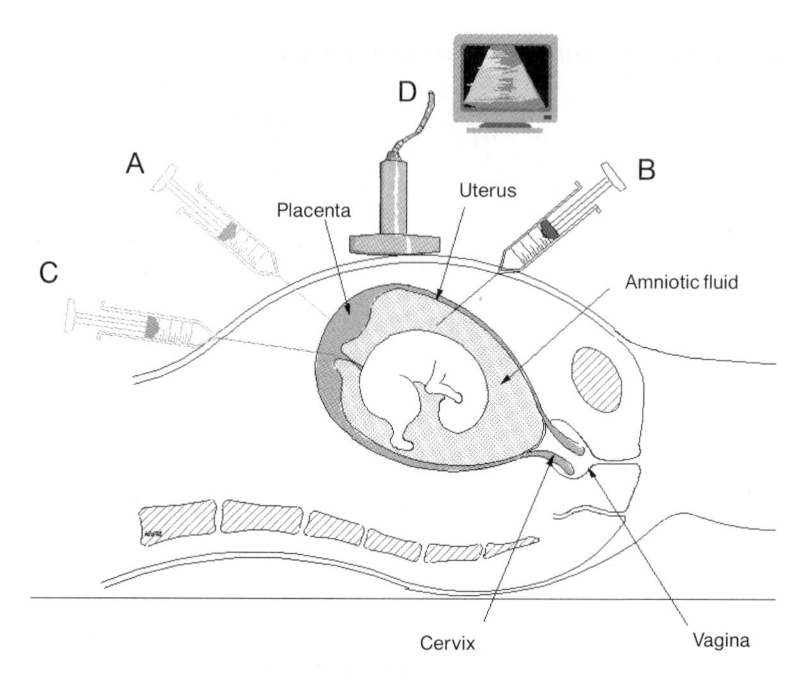

The composition of amniotic fluid varies with gestational age. The foetal cells, many of which are dead, are shed from the skin, lungs and kidneys. Biochemically amniotic fluid resembles foetal urine. The volume increases up to 300 ml by the end of pregnancy.

prenatal testing of some Mendelian disorders. The cells can be grown as a source of foetal DNA although it takes time to culture sufficient cells, which delays the result.

The cells have to be **cultured** for an average of 8–14 days to obtain a foetal karyotype. Occasionally the cells fail to grow and the procedure needs to be repeated. The technique of **fluorescent *in situ* hybridization** FISH (Chapter 6) is being used increasingly in order to produce a more rapid result via the use of chromosome-specific probes on uncultured **interphase** cells. By using probes for chromosomes 13, 18, 21, X and Y, labelled with coloured fluorochromes, the sex of the foetus and the most common viable trisomies can be detected within 24 hours. One disadvantage is that other abnormalities such as **translocations** and **marker** chromosomes will not be detected.

Chorionic villus sampling

Chorionic villus sampling (CVS) involves the removal of a small amount of the placenta. CVS can be done from 11 weeks gestation, which makes it suitable for chromosome analysis following the determination of a high risk of chromosome abnormality as determined by nuchal translucency scans or for known carriers of chromosome abnormalities. The placenta, unlike amniotic fluid, contains spontaneously dividing cells. Two procedures can be used to obtain the foetal karyotype.

Direct preparations take advantage of the spontaneously dividing cells in the outer syncytiotrophoblast layer. It is possible using this technique to obtain a foetal karyotype within 24 hours. However, because of the poor quality of the chromosome preparations available, analysis is limited to the detection of numerical or large structural abnormalities. The tissue can also be **cultured** to provide better quality chromosome preparations and this gives a result in 8–14 days.

The villus tissue is also a rich source of DNA and can be used for molecular diagnosis of some disorders such as cystic fibrosis or Duchenne muscular dystrophy. Some biochemical tests including enzyme assays can also be done.

Foetal blood sampling

Foetal blood sampling (FBS) or **cordocentesis** is the removal of a small amount of foetal blood by inserting a needle into the site where the cord joins the placenta. This is the site where the foetal movement is likely to be least. It is not usually performed before 18 weeks gestation. The blood can be cultured for foetal karyotype analysis and a result obtained in 48 hours. The blood can also be analysed for foetal blood group, infections, etc.

A summarized **comparison** of **prenatal diagnosis techniques** is given below.

- **Amniocentesis:**
  Procedure risk 0.5%.
  Performed in second trimester.
  Widely available.
  Karyotype, some biochemical disorders, DNA (rarely).
- **CVS:**
  Procedure risk 1–2%.
  Performed from first trimester.
  Specialized centres only.
  Rapid karyotype, DNA, biochemical disorders.
- **FBS:**
  Procedure risk 2–3%.
  Performed from second trimester.
  Specialized centres only.
  Rapid karyotype.

Problems associated with prenatal diagnosis

The interpretation of prenatal diagnosis results can sometimes be difficult for several reasons, as described below.

The finding of unexpected chromosome abnormalities

The most common indication for foetal karyotyping is to detect Down syndrome because of advanced maternal age or high risk as determined from biochemical or ultrasound screening. Since 1 in 350 newborn children have a chromosome abnormality, it is not surprising that other karyotypes are also detected at prenatal diagnosis.

Some, such as trisomy 13 and 18, are known to result in severe abnormality. Others, such as sex chromosome abnormalities (Turner syndrome and Klinefelter syndrome, 47,XXY), can have less severe phenotypes and can provide difficult decisions for the parents.

Structural rearrangements may also be found. These may be **balanced** or **unbalanced.** In these cases it is essential to examine the parental chromosomes. Unbalanced rearrangements would be expected to result in foetal abnormality. However, identification of the balanced form of the rearrangement in one parent would be helpful in providing information as to the chromosomes involved and would also indicate a risk of unbalanced offspring in future pregnancies in the couple or any other family member subsequently found to be a carrier.

If the rearrangement were balanced, parental karyotyping would reveal if the rearrangement had arisen *de novo* or was inherited. The presence of a balanced inherited rearrangement in a foetus would be expected to be compatible with normal development. *De novo* rearrangements carry a 5–10% risk of abnormality since it is possible that a small amount of chromosome is missing but not detectable. It is possible that a break (or breaks) had occurred within a gene or that a break (or breaks) had altered the position of the gene, leading to altered expression (see Chapter 3).

Extra structurally abnormal chromosomes (ESACs) are additional chromosomes whose identity is unknown. These can also be inherited without effect. If *de novo* the risk of foetal abnormality can be from 5 to 30% depending on the size and material involved. FISH may be helpful in identifying the chromosome of origin, but if the chromosome is identified it is not possible to know which genes are involved and therefore an accurate prediction of the expected phenotype cannot be given.

The following findings can create **dilemmas in prenatal diagnosis**:

- Sex chromosome abnormalities.
- *De novo* balanced rearrangements.
- ESACs.
- Mosaicism.
- Maternal cell contamination.
- Uniparental disomy.

Other problems of interpretation of prenatal chromosome results

Foetal karyotyping from AF, CVS and FBS is considered to be a reliable technique. However, occasionally the interpretation of results can be difficult because of contamination of the sample with maternal cells or by the presence of mosaicism.

Maternal cell contamination

Maternal cell contamination is usually identified in pregnancies where both male and female cells are found. This is thought to be due to the presence of a male foetus with the female cells being maternal in origin, and will only be noted if the foetus is female when the foetus or the mother has a chromosome abnormality and two different cell lines are discovered. The estimated frequency is 0.5% of AF and CVS cultures. In FBS, testing for the presence of foetal haemoglobin can identify the origin of the blood and hence the presence of any contamination. Maternal tissue contaminating a CVS will not have spontaneous metaphases and will not affect the results from direct preparations.

Maternal cell contamination can also be a problem when undertaking DNA analysis. This may be a particular concern when using the highly sensitive PCR technique. However, direct visual distinction between maternal and villus tissue in the CVS and confirmation of the result on DNA extracted from whole tissue or cultures reduces the chance of error. Alternatively, comparison of foetal and maternal DNA polymorphisms may be helpful.

Mosaicism

Mosaicism is the presence of two or more cell lines with different karyotypes. Mosaicism detected at prenatal diagnosis may have several explanations:

- It may be a genuine reflection of the foetal karyotype (for example, 2–3% of Down syndrome and about 40% of Turner syndrome individuals have mosaic karyotypes).
- It may be due to an error in cell division in culture. The finding of abnormal cells in only one cell culture with a second culture having only normal cells indicates **pseudomosaicism** and is of no clinical significance. This is observed in 1–2% of cell cultures.
- It may reflect mosaicism confined to the placenta, with the foetus having a normal karyotype. This **confined placental mosaicism** (CPM) is detected in 1–2% of CVS.

One particular concern arising from CPM is the possibility that the foetus was originally trisomic and the diploid cells have arisen by loss of one of the extra chromosomes (trisomic rescue). This can result in **uniparental disomy** (UPD), the presence of both chromosome homologues from the same parent. UPD is known to be one of the causes of some abnormal phenotypes such as Prader–Willi syndrome and Angelman syndrome (see Chapter 2).

The discovery of mosaicism may require further tests to help interpret the significance (if any) of the findings. The finding of mosaicism in other tissues confirms the original result. However, the finding of only normal cells does not exclude the presence of an abnormal cell line.

Alternatives to prenatal diagnosis

Some couples cannot contemplate the termination of a pregnancy and therefore prenatal diagnosis is not appropriate unless they wish to know the result in order to take time to prepare for the birth of an affected child. For these couples, alternatives may be more acceptable. The conception of an affected child can be avoided by donation of gametes, either sperm or eggs, or by the identification of a pregnancy as affected prior to implantation. These techniques can be used when one partner is known to carry a genetic disease such as cystic fibrosis or Huntington disease.

Artificial insemination by donor (AID)

AID involves the collection of sperm from an unrelated male donor. The sperm is screened for health risks and the identity of the donor is not revealed to the couple. Attempts are made to match physical features such as colouring, height, etc.

Sperm are placed in the woman's vagina, cervix or womb at the time of ovulation with the aim of achieving natural fertilization and pregnancy. The success rate is about 12% per treatment.

Egg donation

Egg donation is less widely used than AID because of the risks to the donor and the difficulties of the technique. The donated eggs are collected by laparoscopy following hormone stimulation of a female donor with the aim of producing several eggs. These can be fertilized outside the body (**in vitro fertilization**, IVF) or mixed with the sperm and placed together in the Fallopian tubes (gametes in Fallopian tubes, GIFT) to achieve a pregnancy. The success rate is about 20% per cycle.

Preimplantion genetic diagnosis

The availability of IVF has allowed techniques to detect genetic disorders in the embryos prior to replacement into the mother and hence avoid affected pregnancies. Such tests involve the removal of one or two cells at the 3-day blastocyst stage, which are then analysed using FISH, molecular techniques or PCR.

This method has now been used successfully in a number of areas including the diagnosis of foetal sex and single gene disorders such as cystic fibrosis, and in carriers of some chromosome abnormalities. However, the cost, success rate and complexity of the techniques mean that preimplantation genetic diagnosis is likely to be limited to couples with high risk of foetal abnormality.

Future developments

One major concern with current methods of prenatal diagnosis is the loss of normal pregnancies associated with the invasive procedures used. Alternatives that are being investigated include the presence of foetal cells in the maternal circulation. These are known to be present but only in the ratio of one foetal cell to 1.6 million maternal cells. Despite this there have been reports of successful karyotyping, sexing and blood group analysis of foetuses.

Other ways of obtaining test material include washing placental cells from the maternal cervix and sampling the coelomic fluid very early in pregnancy. None of these is as yet a widely used technique.

A second concern is the time taken to obtain a result when testing for Down syndrome and other chromosome abnormalities. The use of FISH is becoming increasingly common, which allows a result to be obtained within 24 hours. The use of PCR techniques on uncultured amniotic fluid samples is being investigated as an alternative method of prenatal diagnosis. It is theoretically possible to detect not just trisomy by gene dosage but also some Mendelian disorders such as cystic fibrosis. The use of multiple assays allows several tests to be performed rapidly on each sample.

There is also research investigating alternatives to increase the sensitivity of the maternal serum biochemistry using other biochemical analytes such as pregnancy associated plasma protein A (PAPP A) and inhibin A and to enable screening to be brought forward to the first trimester.

The ultimate goal, however, has to be to develop **more treatments** for genetic disorders (**gene therapy**) and to identify ways of reducing the incidence of abnormalities as achieved with the folic acid success story.

Suggested further reading

Aitken, D.A. and Crossley, J.A. (1997). Neural tube defects/alpha-fetoprotein/Down's syndrome screening. *Current Opinion in Obstetrics and Gynecology*, **9**, 113–120.

Brock, D.J.H., Rodeck, C.H. and Ferguson Smith, M.A. (eds) (1992). *Prenatal Diagnosis and Screening*. Churchill Livingstone, Edinburgh.

Gardner, R.J.M. and Sutherland, G.R. (eds) (1998). *Chromosome Abnormalities and Genetic Counselling*, 2nd Edn. Oxford University Press.

Orlandi, F., Damiani, T.W., Krantz, D.A. and Macri, J.N. (1997). First trimester screening for fetal aneuploidy: biochemistry and nuchal translucency. *Ultrasound in Obstetrics and Gynecology*, **10**, 381–386.

Sprigg, A. (1995). Foetal malformations diagnosed antenatally. 1: General principles. *British Journal of Hospital Medicine*, **54**, 387–390.

Sprigg, A. (1995). Foetal malformations diagnosed antenatally. 2: Ultrasound diagnosis of foetal structural anomalies. *British Journal of Hospital Medicine*, **54**, 447–450.

Self-assessment questions

1. What is the detection rate for Down syndrome using:
   (a) maternal age;
   (b) maternal serum biochemistry;
   (c) nuchal translucency scans?
2. What is the risk of a child being born with Down syndrome if the maternal age is:
   (a) 20;
   (b) 40?
3. What factors are combined in the triple test?
4. Explain the following terms:
   (a) MOMs

(b) Specificity
(c) Sensitivity
(d) Detection rate.

5. Match the disorder to the most common method of prenatal diagnosis:
   (a) Down syndrome
   (b) Cystic fibrosis
   (c) Anencephaly
   (d) Duchenne muscular dystrophy
   (e) Tay–Sachs disease.
   (i) DNA analysis on CVS
   (ii) Maternal serum AFP and ultrasound
   (iii) Biochemical assay of amniotic fluid
   (iv) Foetal sex and DNA analysis of males on CVS
   (v) Karyotyping of amniotic fluid.

Key Concepts and Facts

Prenatal Diagnosis

- Prenatal diagnosis is the detection or exclusion of abnormality in the foetus during pregnancy.
- The criteria to be met before prenatal diagnosis is considered are that the disorder should be severe, a reliable test should be available, no treatment is available or there are benefits in early diagnosis allowing treatment.

Pregnancies at Risk

- Pregnancies at risk of foetal abnormality can be identified by factors including: maternal age, maternal serum biochemistry, ultrasound scanning, previous child with a Mendelian disorder, family history of a Mendelian disorder, previous child with a chromosome abnormality, parent carries chromosome abnormality, carriers of autosomal recessive disorders identified by population screening.

Methods

- The methods of prenatal diagnosis are amniocentesis, chorionic villus sampling and foetal blood sampling.
- Problems with the methods of prenatal diagnosis include the finding of an unexpected chromosome abnormality, contamination with maternal cells, mosaicism and uniparental disomy.

Alternatives

- Alternatives to prenatal diagnosis include artificial insemination by donor, egg donation and preimplantation genetic diagnosis.

Genetic Counselling

- Genetic counselling has an essential part to play in ensuring that couples are fully informed of the implications of any abnormality found during pregnancy and the risks in future pregnancies to themselves, their offspring and other family members.

Chapter 10
Genetic counselling

Learning objectives

After studying this chapter you should confidently be able to:

Outline the nature of genetic counselling.

Explain the process of genetic counselling.

Be aware of some of the complex issues raised by genetic counselling including genetic testing of children, predictive testing, and the concerns of the effect on insurance and confidentiality.

Genetic disorders have an **impact** in several ways:

- A chromosome abnormality is present in at least 50% of all recognized first trimester miscarriages.
- Of all neonates, 2–3% have at least one major congenital abnormality, which has often been caused by genetic factors.
- Two per cent of all neonates have a chromosome abnormality or a single gene disorder.
- Genetic disorders and congenital malformations together account for 30% of all childhood hospital admissions and 40–50% of all childhood deaths.
- Approximately 1% of all adult malignancy is directly due to genetic factors, and 10% of common cancers such as breast, colon and ovary have a strong genetic component.
- By age 25 years, 5% of the population will have a disorder in which genetic factors play a part.

Introduction

We have seen in the preceding chapters the many ways in which mutations in DNA and also chromosome abnormalities can result in genetic disease, and the ways in which these can be detected using various molecular and cytogenetics techniques. As more and more environmental diseases are successfully controlled, those that are wholly or partly genetically determined are becoming more important. Genetic disorders place a considerable health, emotional and economic burden on affected people, their families and on the community.

In the specialist field of genetic counselling, medically qualified and other health care professionals explain to patients and their families the complex issues involved in the particular genetic disorder, risks and patterns of inheritance, and also organize appropriate laboratory and other diagnostic tests.

Our understanding of how our genetic make-up affects our well-being is increasing, as are the opportunities to diagnose and treat disease. Projects such as that to map the entire human genome will almost certainly lead to the identification of genes which confer susceptibility to common disorders of adult life.

This development, along with increasing public awareness of genetic issues, will lead to ever greater demands on genetic services to provide information and support to individuals and their families in whom genetics is known to play an important role.

The detection of some genetic diseases by prenatal diagnosis and screening programmes is now widespread (see Chapter 9), offering parents a way of avoiding the birth of an affected child or in some disorders being aware in advance that the pregnancy is affected.

Genetic disorders affect not just individuals but also their families. More and more people at different life stages now need access to high quality genetics services. Many ethical issues are involved, ranging from confidentiality (including the effect of information on the insurance industry), to the fear of producing 'designer' babies on demand and eugenics.

This chapter will attempt briefly to discuss the aims of genetic counselling in helping families with genetic disease and will raise some of the ethical issues that need to be considered. It is not possible to cover such a wide area in any detail but only to outline some of the current areas of concern. There are no easy answers.

General principles relating to any aspect of patient contact, including genetic counselling, are:

- **Informed consent:** A patient is entitled to an honest and full explanation before any procedure is undertaken, including details of risks, limitations and implications of each procedure.
- **Informed choice:** A patient is entitled to full information about all options available without duress, including the potential consequences of results to themselves and their families.
- **Autonomy:** The patient is in charge and at any stage can decide to proceed no further with any investigations.
- **Confidentiality:** Which can only be breached in extreme circumstances.

What is genetic counselling?

Genetic counselling is the communication of information and advice about inherited conditions. The person seeking such advice is called the consultand.

The counselling should be:

- **Non-directive**, with no attempt to direct the consultand along a particular course of action.
- **Non-judgemental**, even if a decision is reached which seems ill-advised or is contrary to the counsellor's own beliefs; it is the consultand who has to live with the decision.
- **Supportive**, taking into account the complex psychological and emotional factors that go with the process.
- **Confidential**, although this can raise difficult issues, as will be discussed later.

It should allow those who seek the information to be able to reach their own fully informed decisions without undue pressure or stress.

There are several **reasons for referral** to genetics services:

- Common conditions which may be genetic, e.g. breast and colon cancer.
- Unexplained developmental delay or disability.
- Diagnosis of a genetic disorder after miscarriage or at birth.
- Testing carrier status of family members for Mendelian disorders.
- Genetic management of high risk pregnancies.
- Interpretation of abnormal prenatal results.

The role of genetic counselling is to:

- Establish an accurate diagnosis on which to base the counselling.
- Provide information about prognosis and follow-up.
- Provide risks of developing or transmitting the disorder.
- Discuss ways in which the disease can be prevented or ameliorated.
- Support the family in adjusting to the implications of the genetic disease and the consequent decisions that have to be made.

Establishing the diagnosis

A crucial step in any genetic consultation is that of establishing the diagnosis. If this is incorrect, inappropriate and misleading information would be given, with potentially tragic consequences.

This involves taking a family history which is then displayed on a pedigree as shown in Chapter 4. It should include all abortions, stillbirths, infant deaths, multiple marriages and consanguinity.

It will also involve a full examination, which may include further investigations such as molecular genetics and chromosome studies, as well as referral to specialists in other fields such as neurology, cardiology and ophthalmology.

When people are offered a test they should be informed in advance of the purpose of the test, what is involved and what it will indicate. The explanation needs to cover the following points:

- Will it give a definite diagnosis or only provide information about the risk?
- Will other tests need to be done?
- What it will not show – such as other possible conditions or mutations not specifically tested for.
- Implications for insurance or employment, especially in relation to late onset conditions (see later).

Informed consent should be sought before a test is carried out.

Problems with establishing the diagnosis

Many disorders are recognized as showing genetic heterogeneity, i.e. they can be caused by more than one genetic mechanism (see Chapter 4). This can have important implications when trying to establish the mode of inheritance and hence the risks in other family members. Fortunately, molecular genetics techniques are increasingly able to identify specific mutations involved, thereby providing solutions to some of these problems.

It is essential that the geneticist keeps up to date with new developments, and in particular the identification of new mutations which may help to provide a diagnosis where it was not previously possible.

Certain hereditary disorders can show **different patterns of inheritance**:

- Cerebellar ataxia: AD, AR.
- Ichthyosis: AD, AR, XLR.
- Polycystic kidney disease: AD, AR.

Calculating the risk

The calculation of the risk of a genetic disorder requires the knowledge of its genetic basis. For example, does it have a known Mendelian pattern of inheritance or is it known that genetics plays a part in its aetiology?

The calculation of risks is discussed in more detail in Chapter 4.

Discussing the options

Once a diagnosis has been made and the risks of recurrence or occurrence discussed, the counsellor will aim to provide the consultands with the information necessary, including details of choices open to them, so that they may arrive at their own informed decisions.

This may include discussion of:

- Reproductive options, including prenatal testing and the termination or continuation of an affected pregnancy (Chapter 9) and the psychological and social implications for the individual and their family.
- Reproductive alternatives such as contraception, assisted reproduction, preimplantation genetic diagnosis, gamete donation or adoption.
- Options and implications of testing for carrier status.
- Options and implications of testing for late onset disorders.

Genetic testing of children

The decision of parents as to whether or not to have their children tested for a genetic condition known to be in a family and at what age, can be very difficult. There can be no simple or single formula that will be appropriate for every family. The principal ethical concerns about predictive testing of children are the loss of the child's future autonomy as an adult to make his/her own decisions about testing, and the loss of the confidentiality to which an adult would be entitled. One area that needs to be considered is the possible advantages of early testing in conditions such as FAP where screening can be carried out that may help to reduce the onset or severity of the condition.

When a child is diagnosed as having a genetic condition or chromosome abnormality, his or her future reproductive risk should be recorded in their medical notes.

Predictive testing

The use of molecular genetics techniques enables predictive or presymptomatic testing to be offered for disorders such as familial adenomatous polyposis coli and Huntington disease which will not manifest themselves until later in life.

This has raised difficult ethical issues. One advantage may be that in some cases testing allows earlier screening for the disorder and treatment or other preventative measures can be instigated. Dietary management in carriers of familial hypercholesterolaemia can have a beneficial effect and reduce the risk of heart problems. However, for other diseases, where no treatment is available, identification of a carrier can be very traumatic for the family.

A positive test can also have major implications for close relatives who themselves do not wish to be informed of their disease status. 'Diagnosis by proxy' can result from testing for these autosomal dominant disorders. Such a scenario could result, for example, when an individual requests testing for a disorder because his paternal grandfather is known to be affected, but his own father does not wish to know. A negative result in the consultand does not change his father's 1 in 2 risk. However, a positive result would indicate that his father must also be a carrier.

Consanguinity

Most inherited genetic disorders are rare in the general population. However, should relatives have offspring, the chance of the same recessive gene being present is increased (see Table 10.1).

This is a special problem in genetic counselling because of the increased risk of autosomal recessive disorders. The risk of a child from consanguineous parents being homozygous for a recessive gene is 1 in 32 (see Chapter 4). Marriage between first cousins

Table 10.1 *Genetic relationships and the proportion of shared genes*

Degree of genetic relationship	*Proportion of shared genes*
First degree:	
Parent–child	1/2
Siblings	
Second degree:	
Uncle–niece	1/4
Half siblings	
Third degree:	
First cousins	1/8
Fourth degree:	
First cousins once removed	1/16
Fifth degree:	
Second cousins	1/32

generally increases the risk of severe abnormality in offspring to 3–5% compared to the general population.

Marriage between first and second degree relatives is almost universally illegal. Offspring of incestuous relationships are at high risk of severe abnormality, mental retardation and childhood death. Only about half of children born to first degree relatives are normal, which has important implications for counselling should a pregnancy occur.

Paternity

Paternity can now be confirmed or excluded by DNA fingerprinting techniques (see Chapter 7). Disputed paternity is strictly a legal issue and not one for the genetic counsellor. It is not uncommon to discover non-paternity coincidentally during DNA testing of a family to investigate a Mendelian disorder. This information must remain strictly confidential but it may substantially alter the risks to certain family members and is of great importance in subsequent counselling.

Confidentiality

Identification of a carrier of a condition such as a chromosome translocation or an inherited disorder may have implications for other family members and lead to the offer of tests for the extended family. This can raise questions of confidentiality. A carrier would be urged to alert close relatives to the possibility that they too could also be carriers and their children at risk, and genetic counselling arranged.

If a patient, for whatever reason, refuses to allow this information to be disseminated, despite the consequences being explained, then this wish would usually be respected. Professionals may decide to breach confidentiality when the potential harm to the family member of not being informed, or the potential benefit of being informed, outweighs the potential harm to the individual user of confidentiality being broken.

Genetic information should never be disclosed to third parties such as employers or insurance companies without an individual's written consent.

Genetic registers

A genetic register contains information obtained from genetic counselling and production of a family pedigree, and provides a valuable tool for the genetics services.

The aim of such a register is primarily to establish, as completely as possible, all those people at risk of developing or transmitting a particular disorder so that appropriate counselling can be offered at different life stages. It also permits the long-term follow-up of family members. This is important for children at risk who may not need investigating or counselling for many years, for informing families of new information and research and to continue to follow and support people with long-term or late onset conditions. The genetic register is held on computer and subject to the Data Protection Act. No one is included without having given written consent.

Genetic information and the insurance industry

It is now possible for individuals with a family history of some late onset conditions such as Huntington disease to be tested to determine whether or not they are likely to develop the disorder. With increasing scientific understanding of the contribution made by genetics to a wide range of late onset conditions (such as some cancers and diabetes), it may become possible to identify a proportion of the population as being at a greater risk of developing one of these conditions later in life than is the case for the general population.

There is increasing concern that insurance companies would use this information to discriminate against people at risk of developing a genetic disorder. Currently, there is a voluntary code of practice provided by the Association of British Insurers. The applicant for insurance always has the choice whether or not to take genetic tests. The code does not allow insurers to insist that someone takes a genetic test as a condition of offering them insurance. However, when applying for insurance any existing genetic test result must be given to the insurer unless the insurer has said that such information is not required. Any future genetic tests do not need to be given to the insurer after the policy has been issued.

Test results will not automatically mean refusal or increase in premiums. The information provided will always be kept confidential and the insurers will take professional advice on the information received.

Suggested further reading

Gardner, R.J.M. and Sutherland, G.R. (1996). *Chromosome Abnormalities and Genetic Counselling.* Oxford University Press.

Genetics Interest Group (1999). *Genetics? What Has It Got To Do With Me?* (A resource pack produced by the Genetics Interest Group).

Harper, P.S. (1998). *Practical Genetic Counselling,* 5th Edn. Butterworth-Heinemann.

Self-assessment questions

1. What is genetic counselling?
2. What are the roles of genetic counselling?
3. What would you do if your father/child were at risk of carrying a genetic disorder that might affect you?

Key Concepts and Facts

Genetics and Disorders

- Genetics involves most of us at some stage of our lives.
- Increasing numbers of genetic disorders and underlying susceptibility to common disorders are being discovered.

Genetic Counselling

- Genetic counselling is the communication of information and support of families at risk of a genetic disorder. It includes establishing an accurate diagnosis and calculation of risks of recurrence or occurrence in a family member.
- Genetic counselling may be sought for various reasons, for example following prenatal testing, the diagnosis of abnormality in a child, following miscarriage or death, and to test for carrier status.

Ethical Issues

- Important ethical issues are raised, including confidentiality, the right not to be tested, and testing of children.

Appendix
Glossary of disorders

Modes of inheritance

AD = autosomal dominant
AR = autosomal recessive
XLR = X-linked recessive
XLD = X-linked dominant

Methods of analysis

C = cytogenetics
F = FISH
M = molecular genetics

The OMIM (online Mendelian inheritance in man) database is accessible through the Internet at http://www.ncbi.nlm.nih.gov/Omim.

Each Mendelian character is given an MIM number (see number in brackets below 'disorder') which can be used to access details of each genetic disease.

Disorder	*Mode of inheritance*	*Approximate frequency (if known)*	*Chromosome locus/karyotype*	*Key clinical features*	*Method of analysis*	*Mechanism*
Achondroplasia (100800)	AD	1/12 000	4p16.3	Large head, shortening of limb bones leading to dwarfism	M	Mutations in *FGFR3*
Adult polycystic kidney disease (APKD 1) (173900)	AD		16p13	Gradual formation of renal cysts leading to loss of glomerular filtration	M	Gene mutation and conversion
Angelman syndrome (AS) (234400)		1/25 000–1/40 000	15q11-13	Absent speech, severe retardation, inappropriate laughter, jerky movements, seizures	C, F, M	Deletion, UPD, imprinting error, mutation in *UBE3A* gene
Ataxia telangiectasia (AT) (208900)	AR	1/100 000–1/300 000	11q22.3	Ataxia, immunodeficiency and cancer. Hypersensitivity to ionizing radiation	C, M	Mutations in the *ATM* gene result in a kinase deficiency, disrupting cell cycle and signalling
Basal cell naevus syndrome (109400)	AD	1/57 000 40% new mutations	9q22-31	Skin cancer, macrocephaly, short metacarpals, mild MR	M	Mutations in the *PTCH* TS gene
Beckwith–Wiedemann syndrome (130650)	AD	1/15 000	11p15.5	Overgrowth and large tongue, prone to Wilm tumour	C, M	Duplication, genomic imprinting
Bloom syndrome (210900)	AR	Very rare		Immunodeficiency, acute leukaemias	C	Deficiency of helicase in DNA replication
Burkitt lymphoma (113970)	Acquired		8q24	Cancer of the lymphoid glands	C, M	Activation of *MYC* oncogene
Campomelic dysplasia (114290)	AD		17q24-25	Bowing of long bones, large head, males have ambiguous genitalia	M	Mutations in *SOX9*, position effect
Charcot–Marie– Tooth disease (CMT 1A) (118220)	AD	70% of all CMT (1/3000)	17p11.2	Motor and sensory neuropathy	M	Duplication of *PMP22* in 70%, increased gene dosage
Chronic myeloid leukaemia (CML)	Acquired		t(9;22) (q34;q11)	Raised levels of granulocytes, weight loss, tired, night sweats, splenomegaly	C, F	Fusion of *ABL* oncogene on 9 with *BCR* on 22

Congenital adrenal hyperplasia (CA21H) (201910)	AR	1/5000–1/12 000	6p21.3	With 21-hydroxylase deficiency, girls have virilized genitalia due to a deficiency of cortisol and an increase in adrenocortical hormone	M	Deletions and gene conversion of *CYP21*.
Cri du chat syndrome		1/20 000	5p15	High-pitched cat-like cry, mental retardation, microcephaly, round face	C	Deletions
Crouzon syndrome (CFD1) (123500)	AD	$16.5/10^6$	10q26	Premature fusion of the skull bones leading to beak-shaped nose, small jaw, hypertelorism, protruding eyes	M	Mutations in *FGFR2*
Cystic fibrosis (219700)	AR	1/2000	7q31	Defect in chloride channel leading to sticky mucus in lungs, malabsorption of fats, secretion of excessive salt in sweat	M	Mutations in the *CFTR* gene
DiGeorge syndrome (188400)		1/20 000 CATCH 22: 1/4000	22q11	Cardiac defects, abnormal facies, thymic hypoplasia, cleft palate, hypocalcaemia (CATCH 22)	F	Deletion, ?position effect, haploinsufficiency
Down syndrome		1/700	Trisomy 21 (critical region 21q22.3)	Neonatal hypotonia, epicanthic folds, low set ears, single palmar crease, cardiac anomalies, low IQ	C	?Dosage
Duchenne muscular dystrophy (310200)	XLR	1/3500 males	Xp21.2	Progressive muscle wasting leading to wheelchair by early teens and death by 20 years of age	M	Deletions, some duplications
Edwards syndrome		1/3000	Trisomy 18	Mental retardation, round head, rocker-bottom feet, clinodactyly, cardiac abnormalities	C	?Dosage

continued

Table *continued*

Disorder	*Mode of inheritance*	*Approximate frequency (if known)*	*Chromosome locus/karyotype*	*Key clinical features*	*Method of analysis*	*Mechanism*
Familial adenomatous polyposis coli (FAP) (175100)	AD	1/10 000	5q21-22	Polyps of the colon, becoming malignant	M	Tumour suppressor mutations or deletions in the *APC* gene
Fanconi anaemia (FANCA-227650)	AR	1/100 000–1/300 000	Various complementation groups	Pancytopenia, absent radius and thumbs, acute myeloid leukaemia. Sensitivity to alkylating agents	C	Molecular basis unknown. Deficiency of excision repair?
Facioscapulohumeral dystrophy (FSHD) (158900)	AD	1/20 000	4q35	Muscle weakness of the face, scapula, shoulders and upper arms	M	?Position effect resulting in gene silencing
Fragile X (309550)	Unusual X-linked	1/2000–1/4000 males	Xq27.3	Mental retardation, long face, large ears, macro-orchidism	C, M	Expansion and methylation of triplet repeat CGG
Haemochromatosis (235200)	AR	10% white population are carriers	6p21.3	Elevated iron levels. Cirrhosis of liver, diabetes, hypermelanotic pigmentation of skin, heart failure	M	Mutation in *HFE* gene
Hereditary nonpolyposis coli (HNPCC) (1204356)	AD	5–10% of all colorectal cancer	MSH2: 2p21-22 MLH1: 3p21.3	Colon cancer	M	Mutation or deletion of mismatch repair gene
Holoprosencephaly (HPE3) (142945)	AD	1/16 000–1/53 000	7q36	Developmental abnormality of the face leading to hypertelorism, midline clefting, single nostril, cyclopia	M	Mutation in *SHH*
Huntington disease (143100)	AD	1/10 000–1/18 000	4p16.3	Defect in the caudate nucleus of the brain leading to severe motor disturbance. Late onset	M	Expansion of triplet repeat CAG. Gain of function
Incontinentia pigmenti (IP) (308300) (308310)	XLD		?Xp11 (sporadic) ?Xq28 (familial)	Disturbances in skin pigmentation, abnormalities of the eye	M	Unknown. Gene loci not certain

Klinefelter syndrome		1/1000	47,XXY	Mild learning difficulties, tall, some have gynaecomastia, infertile	C	?Dosage due to extra X
Marfan syndrome (154700)	AD		15q21.1	Arachnodactyly, cardiac and eye abnormalities	M	Mutations in fibrillin gene
Multiple endocrine neoplasia type 2A (MEN 2A) (171400)	AD		10q11.2	Medullary thyroid carcinoma, phaeochromocytoma, parathyroid adenomas	M	Mutation in the *RET* oncogene
Miller–Dieker lissencephaly syndrome (247200)			17p13.3	Mental retardation, smooth brain (lissencephaly), mid-face hyperplasia, small mandible, depressed nasal bridge	C, F	Deletion, haploinsufficiency
Myotonic dystrophy (160900)	AD	1/8000	19q13.2-13.3	Loss of muscle tone leading to drooping eyelids and downturned mouth. Abnormal grip	M	Expansion of triplet repeat CTG
Neuroblastoma (256700)		Rare	2q35 (MYCN) 1p36 (LOH)	Embryonal tumour of the thoracic cavity	C, F, M	Activation of the *MYCN* oncogene, with amplification
Neurofibromatosis type 1 (NF1) (162200)	AD		17q11	Café au lait patches, neurofibromata	M	Mutations in 20% of cases
Pallister–Killian syndrome		Rare	iso (12p)	Severe mental retardation, facial dysmorphism, hypertelorism, sparse eyebrows. Tissue specific; tetrasomy 12p mainly in skin fibroblasts	C, F	Tetrasomy 12p, hence ?dosage
Patau syndrome		1/5000	Trisomy 13	Mental retardation, holoprosencephaly, cleft lip, polydactyly, genital and heart defects	C	?Dosage
Phenylketonuria (PKU) (261600)	AR	1/12 000	12q24.1	Mental retardation treatable by diet, light pigmentation, ezcema, epilepsy, odd posture	M	Mutations in the phenylalanine hydroxylase (PAH) gene

continued

Table *continued*

Disorder inheritance	*Mode of frequency*	*Approximate (if known)*	*Chromosome locus/karyotype*	*Key clinical features*	*Method of analysis*	*Mechanism*
Retinoblastoma (180200)	AD		13q14.1-14.2	Embryonal tumour of the eye	C, F, M	Loss of TS gene(s)
Prader–Willi syndrome (PWS) (176270)		1/20 000	15q11-13	Neonatal hypotonia, childhood obesity, hypogonadism, small hands and feet	C, F, M	Deletion, UPD, imprinting error
Triploidy			69,XXX, 69,XXY or 69,XYY	Severe intrauterine growth retardation with small trunk, syndactyly. Molar placenta depending on parental genome	C	Dosage, imprinting
Turner syndrome		1/2500 of female births	45,X and variants	Neck webbing due to foetal oedema, short, primary amenorrhoea, wide-spaced nipples	C	Dosage, haploinsufficiency
WAGR (194070)			11p13	Wilm tumour, aniridia, gonadoblastoma, mental retardation syndrome	C, M	Deletion and position effect on contiguous gene complex
Waardenburg syndrome type 1 (193500)	AD	1–2/100 000 (2–3/100 000 deaf population)	2q35	Deafness, abnormal iris pigmentation, white forelock	M	Mutation in *PAX3* gene, haploinsufficiency, loss of function
Williams syndrome (194050)	AD	1/10 000	7q11.23	Elfin facies, attention deficit disorder with cocktail party manner, congenital heart defects	F	Deletion of elastin gene
Wolf–Hirschhorn syndrome (194190)	AD		4p16.3	Developmental delay, 'Greek warrior helmet' facies, short philtrum, cleft lip	C, F	Deletion, ?*HOX7* involved in phenotype
Xeroderma pigmentosum (*XPA*-278700)	AR	1/250 000	Various complementation groups	Photosensitivity, skin cancers, neurological disorders	M	Deficiency of excision repair

Answers to self-assessment questions

Chapter 1

1. The nucleus and the mitochondria.
2. DNA comprises an antiparallel right-handed double helix with a sugar–phosphate backbone. Phosphate groups are attached to the pentose sugar via the $3'$ and $5'$ carbons.

   The bases attach to the deoxyribose at carbon atom position 1. They comprise two purines (adenine and guanine) and two pyrimidines (cytosine and thymine), which pair A with T (using two hydrogen bonds) and C with G (using three hydrogen bonds). These lie stacked flat at right angles to the helix.
3. Complementarity is the state in which the double helix of the DNA molecule exists, due to the pairing of bases A with T and C with G. Complementarity enables the copying of one strand of the double helix. If the sequence of bases on one strand is known, the other strand will have a complementary sequence of bases, such that the sequence AGGTTCGGAT should have TCCAAGCCTA on the opposing antiparallel strand.
4. Exons are the conserved coding regions of genes. Introns (or intervening sequences) lie between the exons of genes and are less conserved. A mutation may have clinical consequences in an exon, whereas mutations in introns generally result in polymorphisms, which are harmless differences in DNA sequences between individuals. As these are non-coding sequences, there are no clinical manifestations.
5. Alphoid (α) repetitive sequences are found at the centromeres of chromosomes. Telomeres are found at the ends of chromosomes. Triplet repeats are found throughout the genome, but may have a clinical effect if they exceed a certain number and are located near or within genes.
6. G_1 (variable), S (8 hours), G_2 (4 hours), M (1 hour).
7. Mitosis takes place in somatic cells. There is one division, resulting in 46 chromosomes (diploid: $2n$). Meiosis takes place in the germ cells. There are two divisions, resulting in 23 chromosomes (haploid: n).
8. 2q35.

Chapter 2

1. Gene A is a structural gene; the protein contributes to the structure of the red blood cell. Gene B is a controlling gene; its protein (which may be a transcription factor) interacts with the promoter of gene C and helps to switch it on.
2. During the process of transcription, the whole of one DNA strand is copied (including the introns), producing primary transcript RNA. The introns then form loops which are spliced out, thus leaving a much shorter mRNA strand which is a copy of the exons of the DNA.
3. The gene for steroid sulphatase lies in an area which is not totally inactivated like the rest of the inactive X (Xp22.3). As a female has two Xs she will produce approximately twice the amount of enzyme as a male with one active X.
4. The *HOX* or homeobox genes are important in pattern formation. They ensure that the correct type and number of appendages are present with respect to their symmetrical orientation along the spinal column.
5. The sex determining region on the Y, *SRY*.
6. Female, as this is the 'default' state of the embryo in the absence of the Y chromosome. The second X in normal females is not totally inactivated. The remaining active genes must therefore be required to provide a dosage necessary for normal female development, so in their absence a clinical phenotype will result.
7. Somatic recombination is the mechanism responsible for rearranging the genetic subunits responsible for producing immunoglobulins (antibodies). Without somatic recombination, many separate genes would be required to code for the hundreds of thousands of antibodies needed in the immune system.

Chapter 3

1. A missense mutation results in a base change in the DNA which may have a clinical effect if the resulting amino acid is different (or differently charged) to the normal amino acid. A nonsense mutation produces a stop codon in the DNA sequence, resulting in a prematurely terminated (i.e. shorter) protein.
2. A triplet repeat is a repeating run of three bases. Some repeats cause clinical syndromes if more than a critical number are present in an individual. In Huntington disease, more than 37 copies of a CAG repeat produce an affected phenotype. The repeat lies within the gene and is transcribed, so it must interfere with the function of the normal gene.
3. In fragile X, if the CGG repeat is present in more than 200 copies in a male, the FRAXA gene which lies near to the repeats becomes methylated, leading to cessation of transcription and hence translation.
4. Either a position effect, as the gene for Hunter syndrome has

been separated from a controlling element, or skewed X-inactivation.

5. Maternal uniparental disomy of chromosome 14. This indicates that imprinting is occurring on the maternal chromosome 14 and that the active gene/s on the father's 14 are required for normal development.
6. Normally active genes may be silenced by removal to a new position close to a centromere. This is a position effect due to the inactive heterochromatin found in centromeres.
7. *HOX* (homeobox) genes, as they are involved in pattern formation.

Chapter 4

1. Because the recombination fraction is less than 0.5, the genes must be linked. However, the linkage is not very tight (closely linked genes usually have θ of less than 0.05).
2. Autosomal dominant (myotonic dystrophy, Huntington disease), autosomal recessive (cystic fibrosis, phenylketonuria) and X-linked recessive (haemophilia, colour blindness).
3. Two-thirds.
4. One in three.
5. As the father is affected with RP, the gene on his one abnormal X will be passed on to all his daughters, but as he passes his Y on to his son, the son will be unaffected. The daughters will be phenotypically unaffected carriers of RP. Each has a 50% risk of passing on her abnormal X. There will therefore be a 1/4 chance of a normal daughter, a 1/4 chance of a carrier daughter, a 1/4 chance of a normal son and a 1/4 chance of an affected son.
6. This is essentially a Y-linked disorder as the deletion is on the father's Y chromosome (which he passes on to his son). All his male children would inherit the deletion and would themselves be infertile due to azoospermia. All his daughters would inherit his normal X and would not be at risk.
7. Applying Bayes' theorem to the father:

	Carrier:	*Not a carrier:*
Prior risk	1/2	1/2
Conditional risk	15/100	1
Joint risk	15/200	100/200

Final carrier risk 15/115 = 1/7.67

- Therefore his risk of developing HD is about 1/8.
- The patient's chance of inheriting HD (should his father carry the gene) is 1/2.

The final risk is therefore $1/2 \times 1/8 = 1/16$.

Chapter 5

1. (a) Free trisomy 21; (b) an unbalanced Robertsonian 14;21 translocation.
2. Edwards = 47, cri du chat = 46, Turner = 45.
3. ABCDEBA and GFEDCFG.
4. (a) A deletion too small to detect reliably by banding analysis, without resorting to *in situ* hybridization or molecular analysis. (b) A terminal deletion has a single breakpoint and involves loss of the end of a chromosome arm, while an interstitial deletion comprises loss of a segment from within a chromosome arm.
5. Autosomal trisomies, especially trisomy 16, 45,X (Turner syndrome) and triploidy.
6. A reciprocal translocation involves exchange between two arms of different chromosomes, while a Robertsonian fuses two acrocentrics into a single metacentric chromosome, reducing the diploid number from 46 to 45.

Chapter 6

1. A probe is a double-stranded piece of DNA complementary to a region of interest.
2. Microdeletion detection, using alphoid probes to detect numerical abnormalities and painting for structural abnormalities.
3. Prader–Willi syndrome (obesity, small hands and feet), Wolf–Hirschhorn syndrome (mental retardation, 'Greek warrior helmet' facies) and Williams syndrome (cocktail party habit, typical facies).
4. The correct stringency is a balance of temperature and salt concentration such that all loosely matched probe DNA is washed off, leaving only the tightly bound probe. Stringency can be increased by raising temperature or lowering salt concentration. This will wash off more background (the less well hybridized DNA).
5. 10 kb to 1 Mb.
6. D15S10 for Angelman syndrome.
7. Detection of two different wavelengths and three different wavelengths of light respectively.
8. The 22q11 (DiGeorge syndrome) probe would be used, as heart defects are one of the symptoms.

Chapter 7

1. Whole blood, CVS, cultured cells. The cells must be nucleated, as the DNA is derived from the chromosomes in the nucleus.
2. A low percentage gel. These are medium-sized fragments of DNA, so a 0.8% gel would be adequate. Low percentage gels allow most sizes of fragments to travel but do not resolve small

sizes well. High (2–4%) percentage gels are more suitable for small fragments as they can migrate easily and produce a crisp band.

3. Three for the normal sequence and two for the mutant sequence, as the restriction site for Taq 1 is T*CGA.
   No, as it is part of an intron.
   Polymorphisms.
4. AGCCACGAATTCAT 14 bp = A (runs least far)
   CGAATTCAT 9 bp = B
   AGCCT 5 bp = C
   CGAG 4 bp = D (runs furthest).
5. Southern blotting.
   Fragile X or myotonic dystrophy.
6. A probe which hybridizes to a locus lying close to the location of a disease gene of interest.

   The disadvantage is that there is a risk of recombination between the probe and the polymorphism which tracks with the disorder, such that an incorrect prediction is made.

   The advantage is that as long as the probe is tightly linked to a disease locus, recombination is minimal, and the exact location of the gene need not be known. Gene tracking is then possible.
7. The advantage of PCR is that very little DNA is required – usually nanograms compared with several micrograms for Southern blots. If the sequence of the exon is known, PCR primers can be designed to flank the exon, which can then be amplified in a PCR reaction.

Chapter 8

1. Sporadic. Viruses, chemicals and radiation.
2. Growth factors are normally produced in order to initiate appropriate cell growth in a particular cell type or at a specific time in development. An oncogenic mutation may result in a permanent switching on of the growth factor gene, or other inappropriate gene expression, so there is overgrowth and uncontrolled cell proliferation.
3. The *MYC* proto-oncogene on chromosome 8 is translocated next to an active domain on chromosome 14 (the gene for immunoglobulin heavy chain production); *MYC* is therefore activated.
4. The gene for aniridia must lie near to the Wilms' tumour gene locus to produce the WAGR complex. WT is known to result in a deletion of 11p13. If a constitutional del(11p13) is found, the baby has an increased chance of a second somatic hit on the other chromosome 11 at an early age. A tumour suppressor gene is involved.

5. If MDM2 did not degrade p53, the excess p53 would cause a decrease in the amount of BCL2 and an increase in BAX, leading to apoptosis.
6. Both the *RB1* and the *APC* genes can be inherited constitutionally (i.e. are familial). There is almost 100% certainty (especially with the polyps of FAP) that there will be a second sporadic hit, and therefore these genes appear to act in a dominant manner.
7. Mismatch repair genes proof-read DNA such that normally any potential oncogenic changes are repaired. If these genes themselves are mutated, there will be an increased rate of mutation in other genes, thus increasing the risk of other mutagenic changes.

Chapter 9

1. (a) 30%; (b) 60%; (c) 90%.
2. (a) 1 in 1540; (b) 1 in 100.
3. Maternal serum AFP, HCG, oestriol and maternal age.
4. (a) Multiples of the median. A MOM of 1 means that the values are as expected for the gestation, a value of 2 MOM is raised by twice the expected level and a value of 0.7 is 70% of the expected level.
   (b) Specificity is the extent to which a test detects only affected individuals. False positives are unaffected individuals detected.
   (c) Sensitivity is the proportion of cases detected. False negatives are the affected cases that are missed.
   (d) Detection rate is the numbers of abnormals detected.
5. (a) = (v); (b) = (i); (c) = (ii); (d) = (iv); (e) = (iii).

Chapter 10

1. Genetic counselling is the non-directive and non-judgemental communication of information and advice about conditions where genetics is known or believed to play a part.
2. The role of genetic counselling is to:
   - Establish a diagnosis.
   - Provide information.
   - Estimate risks.
   - Discuss methods of prevention/amelioration.
   - Provide support to the family/individual.
3. Only you can answer that – it is an individual decision.

Index